Klaus Roth

NMR-Tomography and -Spectroscopy
in Medicine

An Introduction

With 122 Figures, Some in Color and 12 Tables

Springer-Verlag
Berlin Heidelberg New York Tokyo 1984

Priv.-Doz. Dr. Klaus Roth
Fachbereich Chemie
Institut für Organische Chemie
der Freien Universität Berlin
Takustraße 3, 1000 Berlin 33

Translator: Terry C. Telger, 3054 Vaughan Avenue, Marina, CA 93933/USA

Title of the original German edition:
NMR-Tomographie und -Spektroskopie in der Medizin
© Springer-Verlag Berlin Heidelberg New York Tokyo 1984
ISBN-13:978-3-540-13442-8/0-387-13076-4

ISBN-13:978-3-540-13442-8   e-ISBN-13:978-3-642-69741-8
DOI: 10.1007/978-3-642-69741-8

2121/3140-543210

To Brigitte
for her helpful support during the writing of this book

and to Timmy
for providing welcome interruptions along the way

Preface

Even the earliest applications of nuclear magnetic resonance (NMR) spectroscopy and tomography to medical inquiries, using experimental apparatus that was primitive by today's standards, demonstrated the extraordinary potential of the NMR method. The subsequent rapid advances in this area were due largely to the efforts of commercial manufacturers, who, by improving magnet and computer designs, were able to produce and market instruments having a remarkable image quality. Experimental data from the first systematic studies on the medical uses of NMR leave little doubt that NMR will gain a permanent place in clinical diagnosis.

The clinician, then, is confronted with an entirely new diagnostic modality. Because NMR has been used extensively in chemistry and physics for years, a great many textbooks are already available on the subject. However, the majority of these have been written for the natural scientist who is well versed in mathematics and physics. Assumptions are made and terms are used that would not be appropriate for a medical or biochemical text. The goal of this introduction, therefore, is to discuss the principles of the NMR technique in terms that are meaningful to the medical student and medical professional.

The author has greatly simplified many details relating to the physics of the NMR phenomenon, believing that the loss of technical accuracy is justified in an introductory text of this kind. Particular compromises were made in the chapters dealing with the measurement of relaxation times and with image reconstruction on the basis of two-dimensional NMR spectra. Thus, if some accuracy has been sacrificed in favor of didactic clarity, this was done solely to provide a better basic understanding of this physically complex diagnostic technique. Readers who wish to consult the original literature are referred to the appendix, which provides more detailed information on the rotating coordinate system and the nomenclature of NMR technology. In addition, a list of selected references is given at the end of each chapter.

This book would not have become reality without the support and active cooperation of many friends and colleagues. I thank Dr. J. Mittner for acquainting me with medical aspects of NMR, and especially for his patience in the translation of medical terminology. I thank Dr. W. Fiegler for his interpretation of the tomograms depicted and for his critical review of the practical examples that are presented. I am grateful to Dr. H. Bauer of Dortmund for reading the manuscript and making useful suggestions, and to Dr. G. Holzmann for his drawings of the rotating coordinate system. I wish particularly to acknowledge the help of Dr. N. Ott of Berlin, who contributed many ideas.

Special thanks are due to the following colleagues who provided me with camera-ready illustrations:

Prof. B. Chance (University of Philadelphia)
Dr. H. Friedburg (Universität Freiburg)
Dr. D. G. Gadian (University of Oxford)
Dr. A. Ganssen (Siemens, Erlangen)
Dr. R. E. Gordon (Oxford Research Systems, Oxford)
Priv. Doz. Dr. W. Huk (Universität Erlangen)
Prof. L. Kaufman (Diasonics, San Francisco)
Prof. J. R. Mallard (M&D Technology, Aberdeen)
Dr. P. Morris (National Institute for Medical Research, London)
Dr. H. Post (Bruker, Karlsruhe)
Prof. R. E. Steiner (Hammersmith Hospital, London)
Dr. H. J. Weinmann (Schering, Berlin)
Dr. I. R. Young (Picker, Wembley)
Dipl. Phys. M. Zabel (Bruker, Karlsruhe)
Dr. Ziedses des Plantes (Kliniken Leyden)
Dr. B. H. Zimmermann (Siemens, Erlangen)

Finally, I express thanks to Mr. B. Lewerich and the staff of Springer Verlag for their thoughtful cooperation.

Berlin, May 1984   Klaus Roth

Table of Contents

1   *Potential Medical Applications of the NMR Technique* . . .   1

2   *Basic Principles of NMR Spectroscopy*   4

2.1   The NMR Experiment   4
2.2   The Chemical Shift   9
2.3   Line Width and Relaxation   13
2.4   Acquisition of the NMR Signal   18
2.5   Design of a Whole-Body NMR System   21

3   *NMR Spectroscopy in Intact Biologic Systems*   29

3.1   In Vivo ^{1}H-NMR Spectroscopy   29
3.2   In Vivo ^{13}C-NMR Spectroscopy   29
3.3   In Vivo ^{31}P-NMR Spectroscopy   34

4   *NMR Tomography*   51

4.1   Basic Principles of NMR Imaging   51
4.2   Image Reconstruction   54
4.2.1   Selecting an Image Plane   56
4.2.2   Image Reconstruction by Back Projection   57
4.2.3   Image Reconstruction by Two-dimensional Fourier
        Transformation   58
4.3   Measurement of Relaxation Times   60
4.3.1   Spin-Lattice Relaxation Time T_1   63
4.3.2   Spin-Spin Relaxation Time T_2   65
4.4   Data Acquisition Strategies in NMR Tomography   70
4.5   The NMR Tomogram   73
4.5.1   Tissue Properties Which Affect the Image   73
4.5.2   Pertinent Imaging Parameters   77
4.5.3   Effects of Flow   82
4.5.4   NMR Contrast Media   84
4.5.5   Imaging Time   86
4.5.6   NMR Tomography of Elements Other than Hydrogen . .   87
4.6   Examples of the First Applications of NMR Tomography   89

4.6.1 Head 89
4.6.2 Torso 92
4.7   Health Risks of NMR Tomography 99
4.7.1 Physiologic Effects of Magnetic and Radiofrequency
       Fields 99
4.7.2 Risks from Metallic Implants 100
4.7.3 Safety Recommendations 101

5   Outlook 105

Appendix A The NMR Experiment in the Rotating Coordinate
            System 106
Appendix B Glossary 110

6   Addendum 114

7   Subject Index 121

1 Potential Medical Applications of the NMR Technique

In 1946 F. Bloch and E. M. Purcell conducted the first experiment in nuclear magnetic resonance (NMR), which in 1952 earned them the Nobel Prize. Since then, NMR has evolved from a method of measuring the magnetic properties of atomic nuclei to a powerful analytical tool yielding data on the structure and concentration of molecules. Through the use of modern computer systems, combined with technical advances in magnet design, it has been possible to improve the sensitivity of NMR measurements by several orders of magnitude. As a result, NMR has become suitable for the investigation of substances occurring in the biologic and medical concentration ranges, and it has stimulated particular interest through its application in intact biologic systems. Recent discoveries on the metabolism of healthy and diseased cells using NMR methods demonstrate the great potential of this experimental technique in clinical diagnostic medicine.

Research on the use of NMR in medicine has focused on two major areas. The first involves a technique developed in 1973 by P. C. Lauterbur, in which NMR data are collected and processed to yield a cross-sectional image of the water distribution

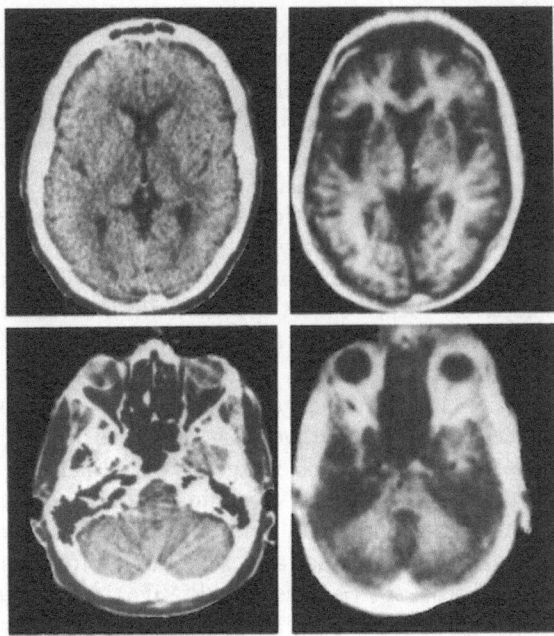

Fig. 1. Axial computed tomograms and NMR tomograms of the human head. The NMR tomograms *(right)* permit a clearer differentiation between gray and white matter in the brain than do the CT images. (Photos from R. E. Steiner and I. A. Young, London)

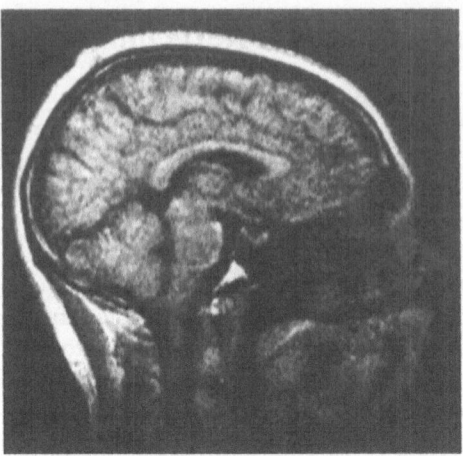

Fig. 2. Mediosagittal NMR tomogram of the head of the author. Unlike computed tomography, in which the image plane is predetermined by the imaging system, the image plane in NMR tomography can be adjusted electronically without moving the patient. In the NMR image, tissues containing water appear light, while air-filled cavities appear dark. The NMR tomogram contains a wealth of detailed anatomic information and is very comprehensive in its imaging capability: soft tissues, bone and intracranial structures are visible on the same image. (Photo from Technicare)

in tissues. By analogy with X-ray computed tomography (CT), this method is called *NMR tomography.*

Figures 1 and 2 show typical NMR tomograms of a healthy human subject taken with prototype NMR instruments. Note that even with these early systems the image quality is outstanding. NMR tomography has already been the subject of many exhaustive studies both in healthy volunteers and in patients with various pathologic conditions.

The second major focus of research is on a technique which complements NMR tomography and was developed by Oxford Instruments, Ltd., in cooperation with a team of Oxford University researchers led by Prof. G. K. Radda. This technique, called *topical magnetic resonance* (TMR), employs a sophisticated instrument setup to selectively record NMR spectra from individual body parts and organs in situ. Phosphorus NMR spectra are of particular value in diagnostic medicine and biology, because they provide a quantitative assessment of adenosine triphosphate, phosphocreatine and inorganic phosphate – the principal energy-storing molecules of the cell. From the relative concentrations of these phosphorus metabolites, much can be deduced about the energetics of living cells.

Figure 3 shows how this technique can be applied to the in-vivo study of oxygen deficiency states in human muscle. Studies with perfused animal organs (liver, kidney, heart) under normal and ischemic conditions clearly show that phosphorus NMR spectroscopy permits a qualitative tissue assessment that is of potential value for organ transplantations and the evaluation of irreversible tissue damage caused by local or global deficiencies of oxygen supply. The results of preliminary diagnostic applications in this area are highly encouraging.

In addition, phosphorus NMR spectra provide a *noninvasive* means of measur-

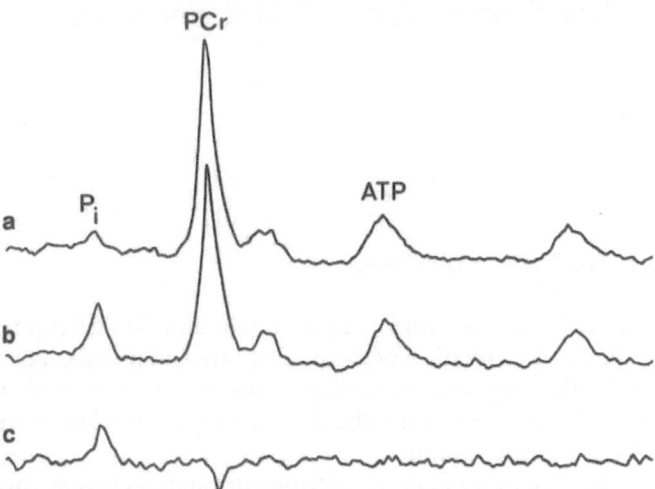

Fig. 3a–c. Phosphorus NMR spectra from a human forearm. The individual signals can be assigned to the phosphorus atoms of three different phosphorus metabolites: phosphocreatine (PCr), adenosine triphosphate (ATP), and inorganic phosphate (P_i). The intensity of each signal is a direct measure of the relative concentration of that compound. **a** Spectrum recorded at rest; **b** spectrum recorded after applying a tourniquet for 10 min; **c** difference spectrum between **a** and **b**. The difference spectrum shows that when an oxygen deficit arose, the concentration of phosphocreatine decreased in favor of inorganic phosphate, while the content of ATP remained constant. The unchanged position of the P_i signal shows that the pH remains constant under ischemic conditions, indicating an absence of ATP synthesis by glycolysis and a failure of lactic acid production. (From documents furnished by Oxford Research Systems)

ing intracellular pH with an accuracy of about 0.1 units. Given the importance of pH as a physiologic parameter, this application alone is of tremendous value. Initial studies of animal organs and human limbs show that continuous intracellular pH measurements by NMR provide an elegant means of diagnosing and localizing enzyme-deficiency diseases, myopathies, and ischemic tissue damage, and of monitoring their response to therapy.

The complementary techniques of NMR tomography and spectroscopy have major advantages over other modalities: They are completely noninvasive, do not expose the patient to ionizing radiation, and, based on current evidence, appear to be harmless. High-risk groups such as pregnant women and newborn infants may be examined as often as desired without risk. The present clinical trials will show whether and in what areas the NMR technique is superior to other established modalities or at least provides a valuable adjunct. Nevertheless, given the capabilities of this method, there seems little doubt that it will be accepted as a routine diagnostic modality in the near future.

2 Basic Principles of NMR Spectroscopy

2.1 The NMR Experiment

In addition to mass and charge, certain atomic nuclei possess a spin, which can be visualized as a rotation of the nucleus around its own axis. This rotation is responsible for the magnetic properties of atomic nuclei on which the NMR experiment is based. Figure 4 illustrates this for the simplest nucleus, that of hydrogen, which consists of a single proton.

Like the movement of electrons through a current-carrying coil, the spin of the positively charged hydrogen nucleus generates an accompanying magnetic field, causing the hydrogen nucleus, like the coil, to behave similar to a magnet. In the absence of an external magnetic field, the nuclear "magnets" are randomly oriented in space. When an external magnetic field is applied, however, the nuclei become oriented either with the field (parallel alignment) or against it (antiparallel alignment), analogous to a compass needle in the earth's magnetic field. Nuclei in the antiparallel alignment are at a higher potential energy level than those in the parallel alignment (Fig. 5).

The energy and population difference between the two alignments (energy levels) is very small. Out of a typical population of 2 million hydrogen nuclei, 999 995 will assume the higher-energy, antiparallel alignment, while the remaining 1 000 005 will be in the lower-energy, parallel alignment. In the NMR experiment (Fig. 6), which is based entirely on this extremely small population difference between ener-

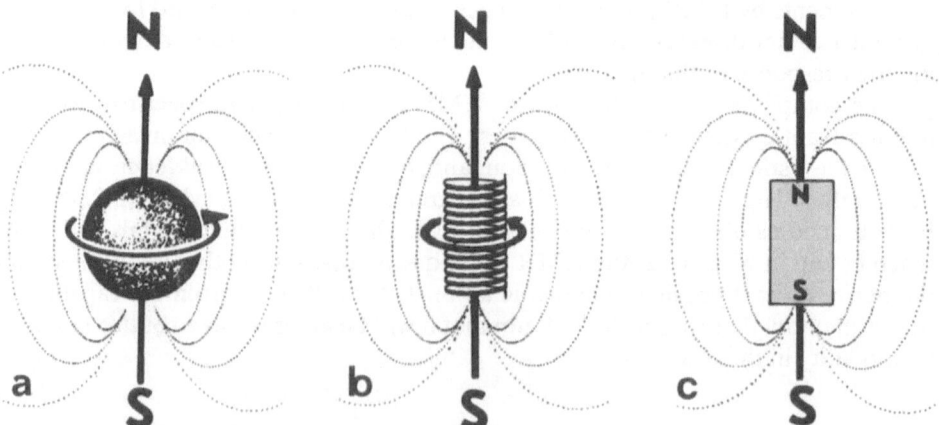

Fig. 4 a–c. The magnetic properties of an atomic nucleus. The spinning nucleus, like any electric charge in motion, generates an accompanying magnetic field (**a**) similar to that produced by a current flowing through a coil (**b**). The fields are not unlike that of an ordinary bar magnet (**c**)

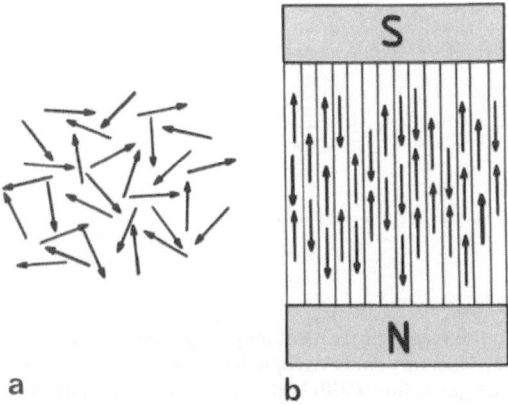

a   b

Fig. 5a, b. Nuclear magnets before and after an external magnetic field is applied. In the absence of an external field (a), the nuclear magnets (symbolized by *arrows*) are randomly oriented in space. When a magnetic field is applied (b), the nuclei become oriented either with the field or against it

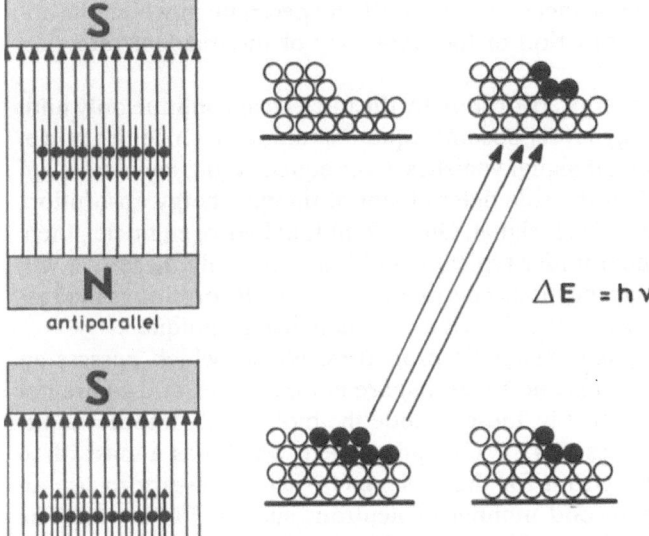

$\Delta E = h \nu$

Fig. 6. The NMR experiment. Nuclei in the parallel alignment are at a lower energy level than those in the antiparallel alignment. By irradiating the sample with radiofrequency energy it is possible to excite transitions from the lower, more densely populated energy level into the higher level. The energy absorption that is associated with this process is recorded as an NMR signal

gy levels, nuclear magnets are made to "flip" from the parallel to the antiparallel alignment. Thus, to generate an NMR signal, the sample to be analyzed is irradiated with electromagnetic energy of frequency ν. When the energy E according to Planck's formula

$$E = h \cdot \nu \quad (h = Planck's\ constant) \tag{1}$$

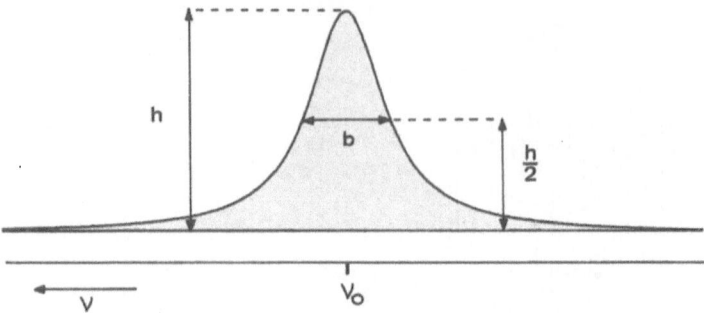

Fig. 7. The NMR signal. The energy absorbed in the NMR experiment may be plotted against irradiation frequency to produce an NMR spectrum. As in any spectroscopic technique, the NMR signal is characterized by resonance frequency ν_0, height h, line width b at half maximum height, and area

precisely equals the energy difference between the two nuclear alignments (resonance), some of the nuclear magnets will reverse their orientation. The energy absorbed during this process can be measured as an NMR spectrum, in which the absorbed energy is plotted as a function of the frequency of the incident radiation (Fig. 7).

When resonance occurs, energy absorption by the nuclei can continue only until the populations of both energy levels become equal (saturation). Once the nuclei are equally distributed, the NMR signal vanishes. Consequently, the *population difference* between energy levels is the sole determinant of the net energy absorption, and thus of the intensity of the NMR signal. Out of 2 million hydrogen nuclei, then, only five will change alignment under typical conditions, and only these five will lead to an observable energy absorption. This extremely weak absorption is the reason for the inherently low sensitivity of NMR as an analytical technique.

All atomic nuclei do not have magnetic properties. Nuclei which possess an even number of nucleons (protons and neutrons) are not magnetic, and so are not accessible to NMR analysis. Notably, these include the biologically important nuclei of carbon, which has a mass of 12 (^{12}C), and oxygen, which has a mass of 16 (^{16}O). To obtain an NMR spectrum for these elements, it is necessary to use an NMR-active isotope having an odd number of neutrons such as ^{13}C or ^{17}O. The magnetic properties and natural abundance of the isotopes of some biologically important elements are shown in Table 1.

Despite the very low efficiency of 5:2000000, the hydrogen nucleus (^1H) is by far the most favorable nucleus for NMR measurements owing to its high relative sensitivity, high tissue concentration, and great natural abundance (99.9%). In other nuclei the energy difference between the parallel and antiparallel states is far smaller, resulting in a lower sensitivity at a given B_0. Also, the relatively low natural abundance of these nuclei may compromise the overall sensitivity of the measurement.

The extremely low sensitivity of NMR spectroscopy can be improved only by increasing the difference between the number of nuclei which populate the different energy levels. One way of doing this is by increasing the strength of the applied magnetic field, thereby amplifying the energy difference between the two align-

Table 1. Properties of some biologically important elements. The nuclei in the bordered rows have no magnetic properties and therefore are NMR-inactive. The relative sensitivity is referred to a sample containing a natural isotope abundance, and the resonance frequencies to a magnetic field strength of 2.3 T, corresponding to 100 MHz for hydrogen

Element	Isotope	Resonance frequency [MHz]	Natural abundance [%]	Relative sensitivity
Hydrogen	^{1}H	100	99.98	100
	^{2}H	15	0.02	0.0002
Carbon	^{12}C	–	98.9	–
	^{13}C	25	1.1	0.02
Nitrogen	^{14}N	7	99.6	0.2
	^{15}N	10	0.4	0.0003
Oxygen	^{16}O	–	99.76	–
	^{17}O	13	0.04	0.002
Fluorine	^{19}F	94	100	85
Sodium	^{23}Na	26	100	13
Magnesium	^{25}Mg	6	10	3
Phosphorus	^{31}P	40	100	8.3
Sulfur	^{32}S	–	95	–
	^{33}S	7	0.76	0.02

ments and producing a greater excess of nuclei in the low-energy, parallel state. The relationship between the magnetic field strength B_0 and the energy difference ΔE is defined by the equation[1]

$$\Delta E = \gamma \cdot \frac{h}{2\pi} \cdot B_0 \qquad (2)$$

where γ is a nucleus-specific constant (the magnetogyric ratio).

Thus, according to Eq. (2), sensitivity can be maximized by using the highest magnetic field strength possible. Equations (1) and (2) can be combined into an expression known as the Larmor relation:

$$\nu_0 = \gamma \cdot \frac{1}{2\pi} \cdot B_0 \qquad (3)$$

This equation describes the simple relationship between the resonance frequency of an atomic nucleus and the strength of the imposed magnetic field. It thus represents the fundamental equation of the NMR experiment.

In whole-body NMR systems, there are limits to how far sensitivity can be enhanced by increasing the magnetic field strength. The radiofrequency radiation that is used to excite nuclear resonance has a limited depth of penetration in biologic material because internal resistance increases with frequency (the "skin effect").

---

1 Strictly speaking, B_0 represents magnetic induction, rather than magnetic field strength, but both quantities are proportional to each other, and this fine distinction is meaningless in the present context. For clarity, then, it is more convenient to regard B_0 as the field strength

The depth of penetration of the radiation is about 25 cm at 10 MHz and 8 cm at 100 MHz. This frequency range may represent the upper limit for whole-body or head examinations.

Figure 8 shows the resonance frequencies of several biologically important atomic nuclei in a magnetic field of 2.3 tesla. The resonance frequencies of the nuclei range between 1 and 500 MHz in magnetic field strengths of 0.1 to 12 tesla. This range, with wavelengths from 1 m to 10 km, is called the radiofrequency (rf) range.

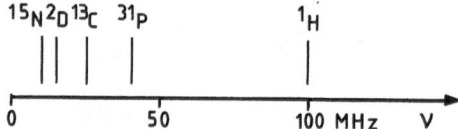

Fig. 8. NMR resonance frequencies of selected nuclei. At a magnetic field strength of 2.3 T, the different nuclei have different resonance frequencies [see Eq. (3)] whose values are determined by a nucleus-specific constant

Frequency [Hz]	Wavelength [m]	Energy [mol⁻¹]	Radiation	Biologic effect	Medical application
10^{20}	10^{-10}	100	γ	Breaking of chemical bonds	γ−radiation therapy
		10	X−ray		Radiography and CT
		1keV			
		100	UV	Excitation of: Electrons	UV irradiation
10^{15}	100	10eV			
			Visible		
	10^{-5}	10		Molecular vibration	
		1	Infrared		
		kcal 100		Molecular rotation	Diathermy
		10	Microwave		
10^{10}		1cal	Radar		
		1J	TV		
	1		VHF (FM)		Short−wave Therapy NMR
			Short wave		
			Medium wave (AM)		
10^{5}			Long wave		
	10^{5}		Ultrasound		Ultrasonography
			Audible sound		

Fig. 9. Overview of the electromagnetic spectrum

The position of these waves within of the electromagnetic spectrum is shown in Fig. 9. The radio waves used in the NMR experiment are some 10^{-11} times less energetic than the X-rays commonly used in human medicine. Damaging effects from these radio waves are, therefore extremely unlikely.

2.2 The Chemical Shift

High-resolution NMR spectra show that hydrogen atoms associated with different kinds of chemical bonds have different resonance frequencies, and thus produce different NMR signals. This is due basically to the different electron environments of these hydrogen atoms. Figure 10 compares the hypothetical spectrum of an isolated nucleus to that of a nucleus surrounded by a full electron shell.

While the applied magnetic field can penetrate freely to an isolated nucleus, the presence of shell electrons creates a magnetic "screen" around the nucleus which weakens the applied field strength. In accordance with the Larmor relation, Eq. (3), different effective magnetic field strengths at the nucleus lead to corresponding

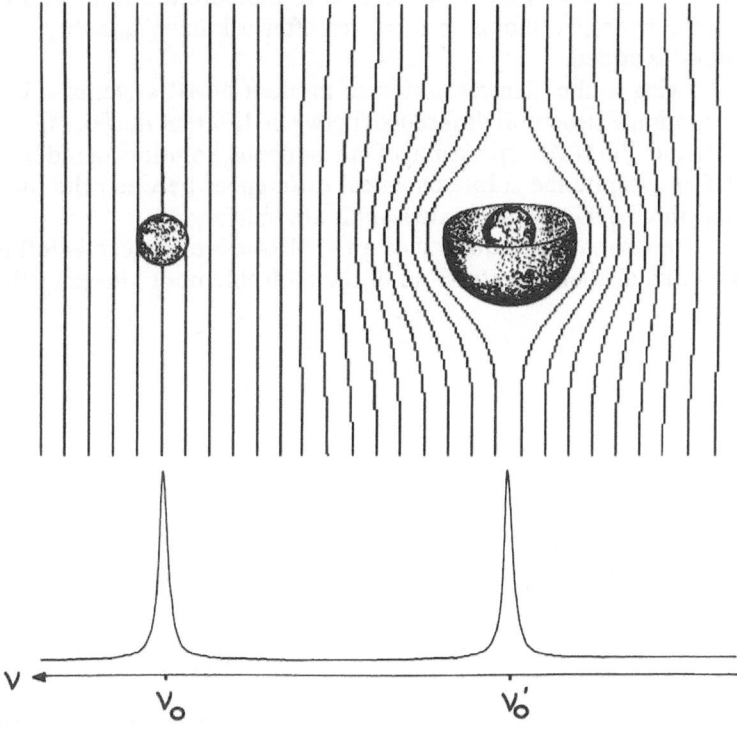

Fig. 10. Effect of the shell electrons of an NMR-active atom on resonance frequency. The electrons in the shell weaken the magnetic field strength that is experienced by the nucleus. This also decreases the resonance frequency from ν_0 for the isolated nucleus to ν_0' for the nucleus with electrons. Because the electron density around the nucleus depends on the chemical structure of the molecule, the resonance frequency of a nucleus is a highly characteristic structural parameter

changes in the resonance frequency. This effect is called the "chemical shift." Because only the *relative* frequency difference between the NMR signals of atoms with different chemical bonds is of importance, the chemical shift may be characterized in terms of the "δ scale," defined as

$$\frac{\nu_0 - \nu_0'}{\nu_0} = \delta \tag{4}$$

At a radiofrequency of 100 MHz, the NMR frequency shifts to the hydrogens caused by chemical bonds are of the order of several 100 Hz, yielding δ values of several 10^{-6}. For practical reasons it is customary to express the δ value in parts per million (e.g., 10^{-6} ppm). As a relative quantity, the δ value must always be referred to a standard signal to which a specified value (usually zero) has been assigned. In principle, any signal may be chosen as the reference, but for most nuclei standard signals have been agreed for comparison.

As a practical example, Fig. 11 shows the ^{1}H-NMR spectrum at 60 MHz of methanol (CH_3OH). Integration of the area below the resonance peaks is proportional to the relative concentration of the hydrogen atoms absorbing energy at that frequency. In the presented case, the measured intensity ratio of 3:1 enables the signals to be assigned to the chemically different hydrogen atoms of methanol. NMR spectroscopy can also be used to measure the concentrations of chemically diverse hydrogen atoms in mixtures after a known quantity of a reference compound is added.

Owing to the high resolution of modern NMR systems, it is possible to detect even minute structural differences between different nuclei. Figure 12 illustrates this with the ^{31}P-NMR spectrum of an aqueous solution of adenosine triphosphate (ATP). Despite the subtle chemical differences between the three phosphorus atoms, three distinctly separate signals are obtained.

The resonance frequency of an NMR-active nucleus is influenced not just by the internal chemical structure of the molecule under investigation, but also by in-

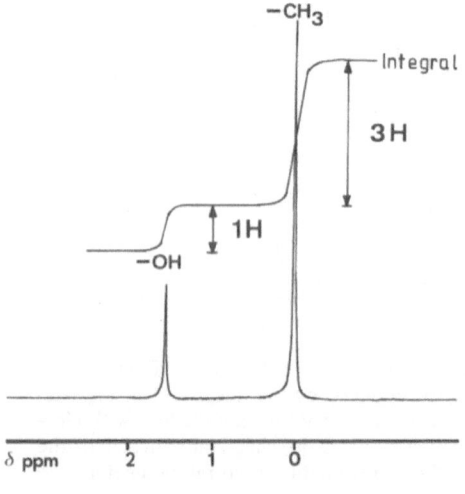

Fig. 11. ^{1}H-NMR spectrum of methanol (CH_3OH). Two separate signals are present in the NMR spectrum. Electronic integration of the signal areas gives an intensity ratio of 3:1 for the chemically different hydrogen atoms in the molecule, corresponding to the three methyl-group protons and one hydroxyl proton of methanol

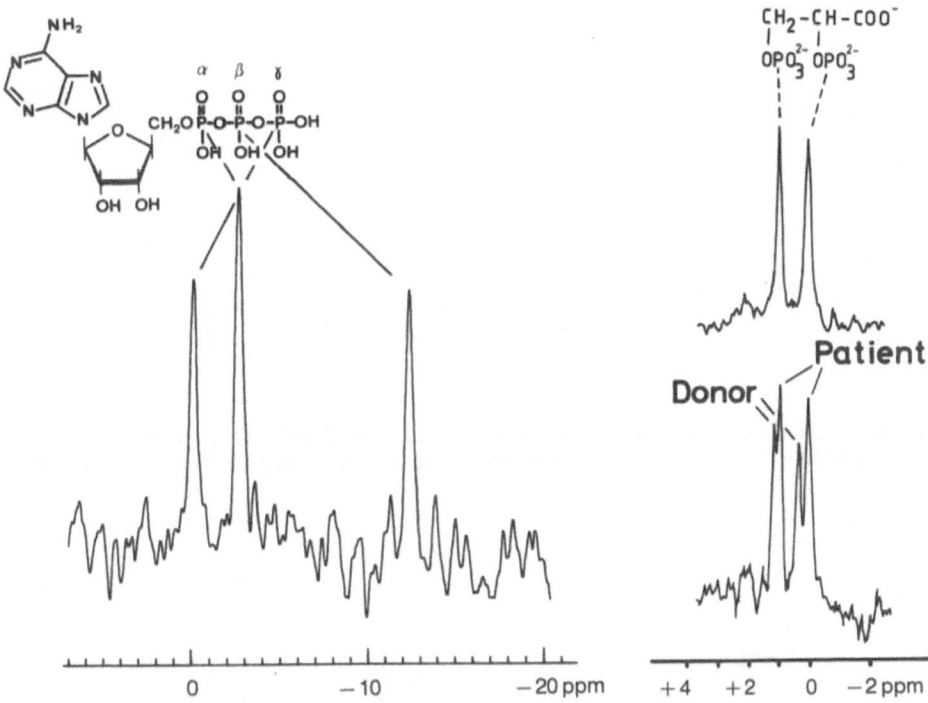

Fig. 12 *(left).* ^{31}P-NMR spectrum of adenosine triphosphate (ATP). The ^{31}P-NMR spectrum of an aqueous solution of ATP shows three separate signals for the phosphorus atoms, which are bound in different chemical environments

Fig. 13 *(right).* ^{31}P-NMR spectra of erythrocytes. The aqueous suspension of erythrocytes from a patient with thalassemia *(top)* shows separate signals for the two chemically diverse phosphorus atoms of 2,3-diphosphoglycerate. Following a blood transfusion, both signals are clearly split *(bottom).* The different chemical shifts of the phosphorus signals from the donor and patient erythrocytes are based on differences in the way the 2,3-diphosphoglycerate is bound to the normal (donor) and fetal (patient) hemoglobin. [Labotka RJ, Honig GR (1980) Am J Hematol 9: 55]

tramolecular bonding. Figure 13 shows the ^{31}P-NMR spectrum of an aqueous suspension of erythrocytes from a patient with thalassemia major. The two peaks correspond to the two phosphorus atoms in the 2 and 3 position of 2,3-diphosphoglycerate. The spectrum of the erythrocytes following a partial blood transfusion shows a doubling of the NMR signals. This slight change in the chemical shift of both phosphorus atoms in the blood of the patient and the donor is due to differences in the bonding of the 2,3-diphosphoglycerate to the normal (donor) and fetal (patient) hemoglobin. Integration of the signals enables one to assess quantitatively the efficacy of the transfusion.

The empirically determined δ values for several characteristic structural elements of the biologically important nuclei of ^{1}H, ^{13}C, and ^{31}P are shown in Figs. 14–16.

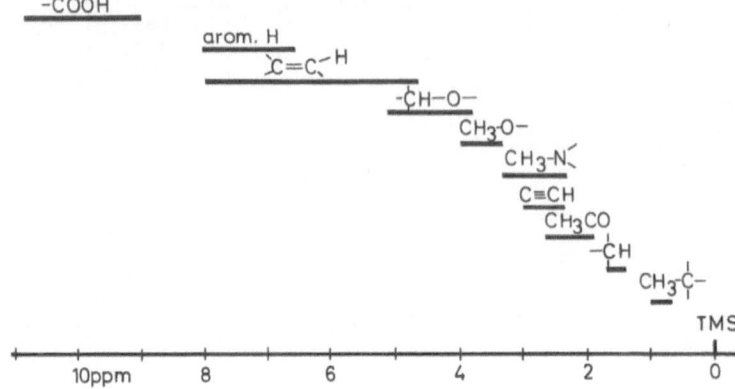

Fig. 14. ^{1}H chemical shifts of typical structural elements. The δ scale of the chemical shifts of hydrogen nuclei (= protons) is customarily referred to tetramethylsilane [(CH$_3$)$_4$ Si = TMS] as zero

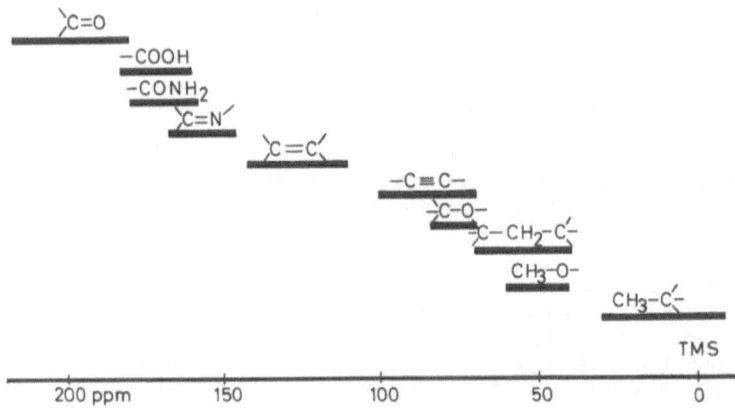

Fig. 15. ^{13}C chemical shifts of typical structural elements. As in the proton shifts, tetramethylsilane (TMS) serves as the zero reference for the position of the ^{13}C-NMR signals

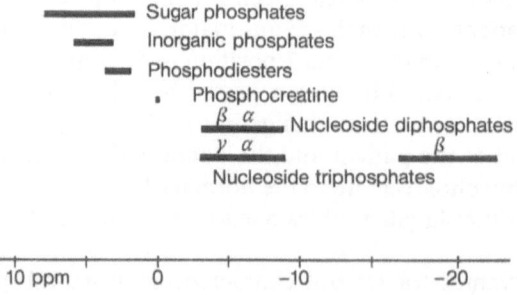

Fig. 16. ^{31}P-NMR chemical shifts of some biologically relevant structural elements. For in vivo measurements, the ^{31}P-NMR signal of phosphocreatine usually serves as the zero reference for the δ scale

2.3 Line Width and Relaxation

Due to the inherently low sensitivity of the NMR technique, NMR signals can be obscured by background noise. If the concentrations of the molecules of interest are very low, the NMR signals will be completely masked by noise. However, it is possible to enhance the NMR signals by summing individual scans (up to 100 000 or more) with a computer. The ratio of the signal intensity to the noise intensity (signal-to-noise ratio, S/N) will increase as the square root of the number of scans, according to the formula

$$\left(\frac{S}{N}\right)_n = \sqrt{n}\left(\frac{S}{N}\right)_1 \tag{5}$$

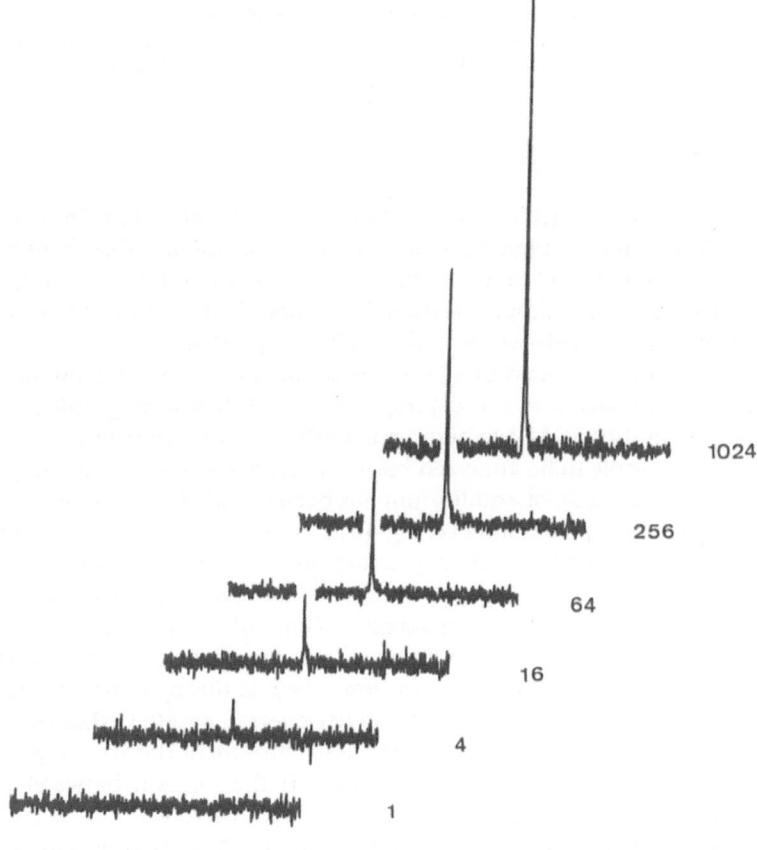

Fig. 17. Enhancement of spectral quality by summation. The signal-to-noise ratio can be improved by summing multiple individual spectra in a computer. In the present case of a 0.4 mmol aqueous solution of sodium succinate ($NaOOCCH_2CH_2COONa$), the hydrogen NMR signal after one scan is indistinguishable from noise. Spectral quality can be greatly enhanced by making successive scans of the NMR spectrum, each fourfold increase in the number of scans bringing a twofold improvement in the S/N ratio

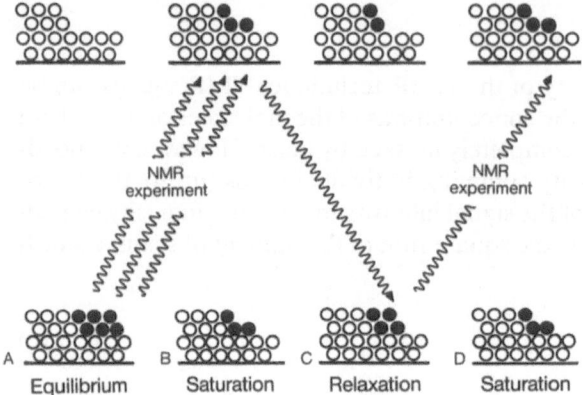

Fig. 18. Effect of repetition rate of NMR experiments on signal intensity. In the first NMR experiment, irradiating the sample with radiofrequency energy produces a transitory saturation of both energy levels (*B*). This is followed by a relaxation process in which energy is lost to the environment, and the nuclei return to the lower energy level (*C*). If a second NMR experiment is done before equilibrium (*A*) is restored, the smaller population difference that exists (*C*) will cause a decrease in signal intensity

where $(S/N)_1$ and $(S/N)_n$ represent the S/N ratio after 1 and n scans respectively.

Thus, while a single-scan spectrum is indistinguishable from noise, the averaging of spectrum and noise enables the distinction to be made (Fig. 17).

However, this simple relationship is not always valid. When series of scans are run in *rapid succession,* deviations from Eq. (5) will occur; the total intensity becomes a complex function of the repetition rate. To understand the basis of this phenomenon, which is of critical importance in NMR tomography, let us take a closer look at multiple NMR experiments performed in rapid succession (Fig. 18).

The sample to be analyzed has been placed in an external magnetic field, and a characteristic state of equilibrium has been established between the parallel and antiparallel nuclear alignments (Fig. 18a). If the sample is now irradiated with rf energy at the resonance frequency, some nuclei in the parallel alignment will absorb energy and tip into the higher-energy, antiparallel alignment, until a uniform distribution (saturation) has been achieved (Fig. 18b). This state is not permanent, however, and when the rf energy is turned off, the nuclei tend to redistribute themselves in a manner that restores equilibrium. They do this by returning from the high-energy, antiparallel alignment to the lower-energy, parallel alignment, simultaneously releasing their absorbed energy to the environment (lattice) (Fig. 18c). This process is called "relaxation," and takes time. If the interval between successive NMR experiments (repetition time, T_R) is too short, there will not be enough time for the nuclei to return to equilibrium (Fig. 18d). The decreased population difference that exists at this time necessarily leads to a smaller energy absorption when radiofrequency energy is reapplied (Fig. 18d), resulting in a decreased signal intensity in the spectrum. The relaxation process has a characteristic time constant called "the spin-lattice relaxation time T_1." A simple exponential relationship describes the increase in signal intensity and the repetition time T_R (Fig. 19):

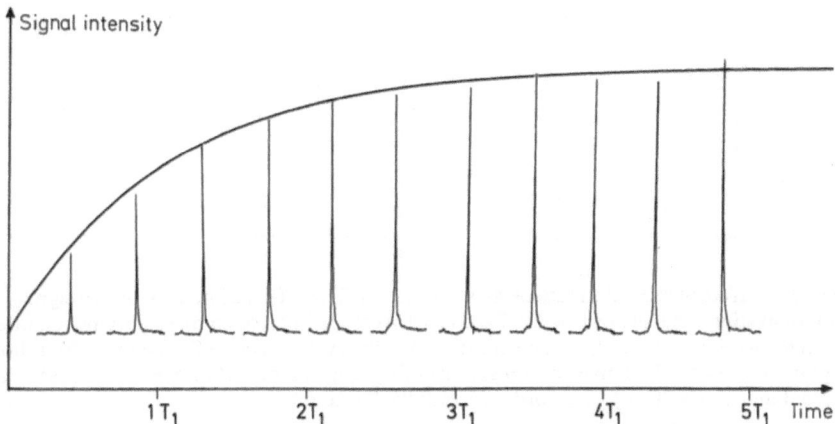

Fig. 19. Time course of signal intensity. After an NMR experiment is performed, some time is required for the nuclear magnets to lose energy and regain their original distribution. As a result of this relaxation process, which is characterized by the time constant T_1, the signal intensity measured in a second NMR experiment increases as a function of time. After an interval of $3 \cdot T_1$ the signal has already attained 95% of its maximum intensity

$$J = J_0 \cdot \left[1 - \exp\left(-\frac{T_R}{T_1} \right) \right]. \tag{6}$$

Since the first NMR experiment at $T_R = 0$ leads to a uniform distribution of the nuclear magnets between both energy levels, no further energy absorption would be possible if a second NMR experiment were done right away. During the process of spin-lattice relaxation, nuclear magnets make a transition from the higher-energy to the lower-energy state, with the difference in energies being transferred to the molecular environment (lattice). The longer the delay before rf energy is reapplied (i. e., the further spin-lattice relaxation is allowed to progress), the greater will be the observed intensity of the second NMR signal (Fig. 19).

Owing to the recovery of signal intensity after saturation, the term "saturation-recovery experiment" is also applied to multiple resonance experiments that are performed in a rapid succession.

To calculate the total increase in the S/N ratio with the number of scans of the NMR spectrum, it is necessary to modify Eq. (5) as follows:

$$\left(\frac{S}{N} \right)_n = \sqrt{n} \cdot \left(\frac{S}{N} \right)_1 \cdot \left[1 - \exp\left(-\frac{T_R}{T_1} \right) \right]. \tag{7}$$

Equations (7) and (5) are equivalent only when the interval T_R between successive NMR experiments is greater than $3T_1$.

Like the δ value of the chemical shift, the relaxation time T_1 of a nucleus within a molecule is a characteristic quantity. Figure 20 shows the intensity of two signals with different relaxation times as a function of T_R. By adjusting T_R, it is possible to vary the two signal intensities within certain limits.

The T_1 value of a nucleus in a molecule depends not only on the structure of the molecule itself, but also on its mobility relative to its surroundings (lattice). Thus,

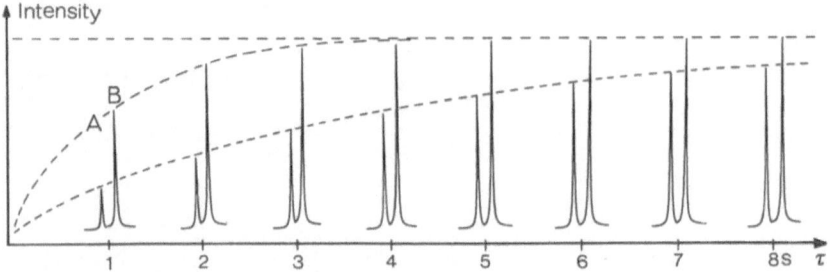

Fig. 20. Effect of repetition rate on signal intensity. The relative intensity of two signals with different relaxation times [T_1 $(A) = 4s$; T_1 $(B) = 1$ s] can be influenced by modifying the interval T_R between successive scans of the spectrum. If this interval exceeds $3T_1$, the intensity ratio will correspond to the actual relative concentrations of hydrogen atoms A and B. With shorter T_R intervals, signal intensity will vary in accordance with the T_1 values

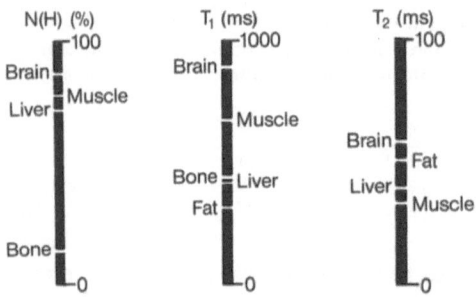

Fig. 21. Water content N(H), spin-lattice relaxation time T_1, and spin-spin relaxation time T_2 of various human tissues. (After Buchmann and Heinzerling (1983) GIT Lab Med 6: 102)

the T_1 value provides information on the interaction between the molecule and its environment. In pure water, for example, hydrogen nuclei have a T_1 value of approximately 3 s. By contrast, the hydrogen nuclei in tissue water have T_1 values ranging between 0.2 and 0.8 s, depending on the type of tissue involved (Fig. 21).

Although the differences in relaxation time of the hydrogen in various body tissues are not yet fully understood, the T_1 value appears to depend on the nature and strength of the bond which holds the water to cellular proteins. A hydrogen nucleus in a water molecule that is relatively firmly bound to the surface of a cellular protein can transfer excess energy to the environment much more easily (short T_1) than a hydrogen nucleus in a free water molecule.

With NMR tomography, it is possible to differentiate between tissues which contain an equal amount of water but have different T_1 values by adjusting the interval T_R between successive scans of the spectrum. If this interval exceeds $3T_1$, the signal intensities correspond to the water content, which is equal in both tissues, and no discrimination is possible. However, if a faster scanning rate is employed (smaller T_R), the different relaxation times of the tissues cause the water in both tissues to give signals of different intensities (Fig. 22).

Besides the spin-lattice relaxation time T_1, the return to equilibrium is characterized by yet another relaxation mechanism, called "spin-spin relaxation," which is

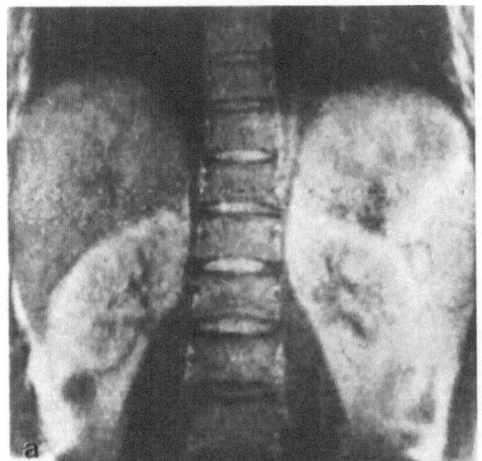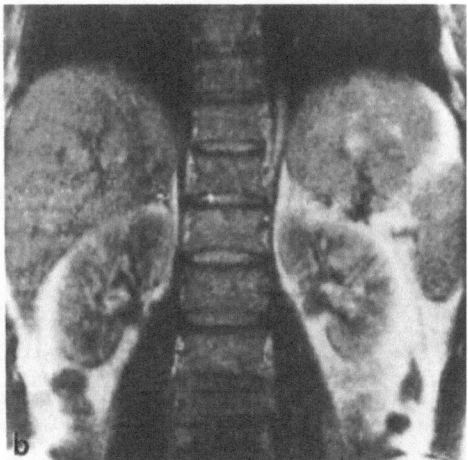

Fig. 22 a, b. Coronary NMR tomograms of the retroperitoneum. The effect of repetition time is reflected in the degree of image contrast that is obtained. By a suitable choice of imaging parameters, T_1 variations can be exploited to discriminate between tissues having the same water content. **a** The interval T_R between successive scans was 1600 ms, with a total imaging time of 13 min. Owing to the long delay, the image brightness corresponds essentially to the water content of the tissues. In **b**, the scan interval was only 402 ms, and the total imaging time was shortened to 7 min. Under these conditions the NMR tomogram is determined both by the water content and the T_1 relaxation time of the tissues, providing a better overall tissue discrimination. Both kidneys, the perirenal fat, the psoas muscle, the crura of the diaphragm, the liver, spleen, and basal lung regions are well visualized

independent of T_1 and is represented by the time constant T_2. Spin-spin relaxation is due to the alterations in the magnetic field strength experienced by a nucleus in the presence of neighboring molecules. Whereas free water molecules always experience a constant field strength owing to the rapid, random motion of their neighbors, the molecules in the vicinity of surface-bound water molecules are fixed, and so their local magnetic fields do not average to zero. As a result, the NMR signal from the same atom may appear at slightly different resonance frequencies in two chemically identical molecules, depending on the magnitude of local field strength variations. Averaged over all the molecules in a sample, this leads to a broadening of the line width according to the formula

$$b = \frac{1}{\pi T_2} \tag{8}$$

where b is the line width at half the maximum height of the resonance peak.

In liquids and solutions, T_1 and T_2 are approximately equal, whereas in molecules that are adherent to surfaces or bound to cellular proteins T_2 appears shorter (Fig. 23). In solids, T_2 is so short that the corresponding NMR signals are often indistinguishable from the baseline owing to their very broad line widths. Thus, NMR signals from solids (e. g., bone) or from large molecules such as enzymes, proteins, membranes, or membrane-bound molecules *cannot* be observed using the NMR methods described in this survey. *In the NMR spectra from biologic material, signals come exclusively from small, highly mobile molecules (tissue water, ATP, etc.).*

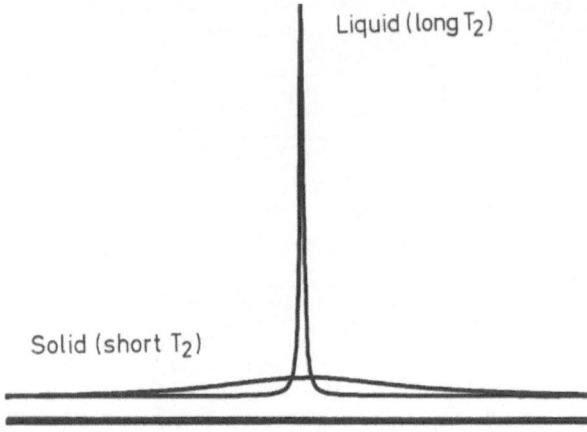

Liquid (long T₂)

Solid (short T₂)

Fig. 23. Signal width and relaxation time T_2. The spin-spin relaxation time T_2 is an indirect measure of the mobility of the resonating nucleus. Solids have a short relaxation time, leading to an increase in line width

Table 2. Relaxation times T_1 (in seconds) of tissue water in various normal and tumorous human tissues

Tissue	Normal	Tumorous	Tissue	Normal	Tumorous
Skin	0.62	1.05	Esophagus	0.80	1.10
Skeletal muscle	1.02	1.41	Stomach	0.76	1.23
Spleen	0.70	1.13	Bowel	0.64	1.22
Lung	0.79	1.11	Liver	0.57	0.83
Bone	0.55	1.03			

From a diagnostic standpoint, it is important to note that the relaxation times of tissues may be measurably altered by the presence of pathologic changes (Table 2). This change in relaxation time forms the basis for tissue discrimination in NMR imaging.

2.4 Acquisition of the NMR Signal

In contrast to roentgenography and spectroscopic methods, an NMR signal cannot be acquired directly simply by measuring the attenuation of the applied radiofrequency energy, because the degree of energy absorption is far too small. An indirect approach must therefore be used.

A very short burst of radiofrequency energy, generally lasting several microseconds, is applied by means of a coil that is wound about the sample. This burst of radiofrequency energy is called a *pulse*. The coil is tuned to an energy distribution that will induce resonance simultaneously in all susceptible nuclei in the sample (Fig. 24).

Under optimal conditions, all energy levels will be equally populated after the pulse corresponding to a maximum degree of energy absorption (saturation). After the pulse, the nuclei return to equilibrium by relaxing back to the lower energy state, releasing excess energy to the environment (see Fig. 18). This process can be detect-

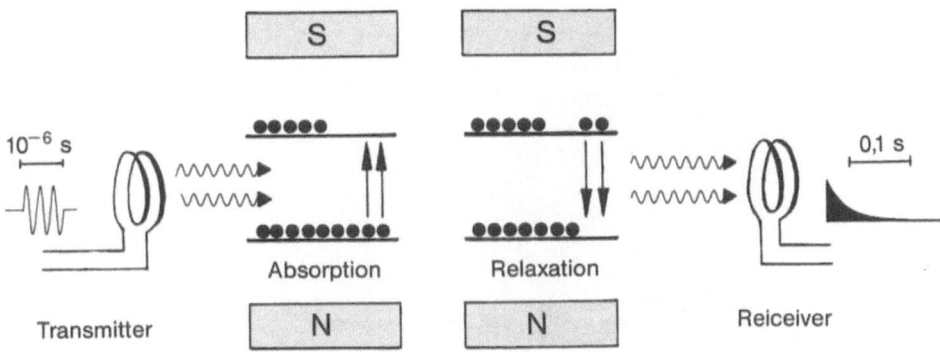

Fig. 24. NMR pulse experiment. A short radiofrequency pulse is applied to saturate both energy levels in the sample. Afterward, the nuclei return from the upper energy level to the lower energy level, emitting radiofrequency energy that can be recorded as a decay curve

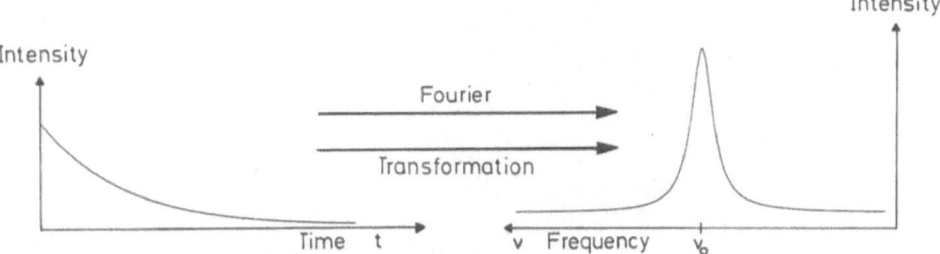

Fig. 25. The Fourier transformation is a mathematical operation by which a function can be converted from the time domain to the frequency domain. In the simplest case, the decay of energy emission is described by a simple exponential curve, which can be converted to a normal NMR signal by Fourier transformation

ed and measured by means of a receiver coil. Thus, in modern NMR spectrometers, an NMR signal is acquired not by measuring energy absorption *during* resonance, but by measuring the *subsequent release* of the absorbed energy during relaxation. Because radiofrequency transmission and reception take place at different times, a single radiofrequency coil may be utilized both as transmitter and receiver.

The signal received by the coil has the shape of a decaying exponential function:

$$I = I_0 \exp\left(-\frac{t}{T_2}\right) \tag{9}$$

where T_2 is the spin-spin relaxation time described in Section 2.3.

If several susceptible nuclei are present in the sample, the measured decay curve (called the "free induction decay") does not have a simple exponential shape like that in Eq. (9), but represents the sum of many different decay curves. This curve is no longer amenable to direct analysis, especially when different resonance frequencies are involved. It is necessary to convert the decay curve, representing the change in signal intensity as a function of time, into a conventional NMR spectrum which represents signal intensity as a function of frequency. This conversion from the time domain to the frequency domain is done by means of the Fourier transformation –

Fig. 26. Jean Baptiste Joseph Fourier was a man of many talents. Born the son of a tailor in Auxerre, France, in 1768, Fourier first worked as a teacher in his home town. During the French Revolution he became involved in local politics. His actions on behalf of victims of the Reign of Terror led to his arrest, and despite direct appeals to Robespierre, Fourier was sentenced to death. On July 28, 1794, the day before the execution, Robespierre was overthrown, and Fourier was granted a pardon. He accompanied Napoleon on his Egyptian campaign and probably served as governor of Lower Egypt. Following his return to France, he was named Prefect of the Department Isére centered at Grenoble by Napoleon, who recognized his diplomatic and organizational skills. In this office Fourier performed many useful services, including the reclamation of large land areas around Bourgoin. Following Napoleon's exile, Fourier fell into disfavor and earned a living as Director of the Bureau of Statistics. Initially, Louis XVIII refused Fourier admission into the Adadémie des Sciences for political reasons, but in 1817, on the strength of pleas from fellow scientists, Fourier was admitted to the Académie, and in 1823 he became its secretary. In 1827 he was accepted as a member of the Académie Française. He died in Paris in 1830. It was during a theoretical analysis of heat conductivity that Fourier developed the mathematical process which bears his name. He began work on the Fourier transformation while still in Egypt and completed it in Grenoble. In 1822 Fourier published his results in his famous work, *Theorie Analytique de la Chaleur.* (Biographical data from D. Shaw; illustration from Austrian National Library, Vienna)

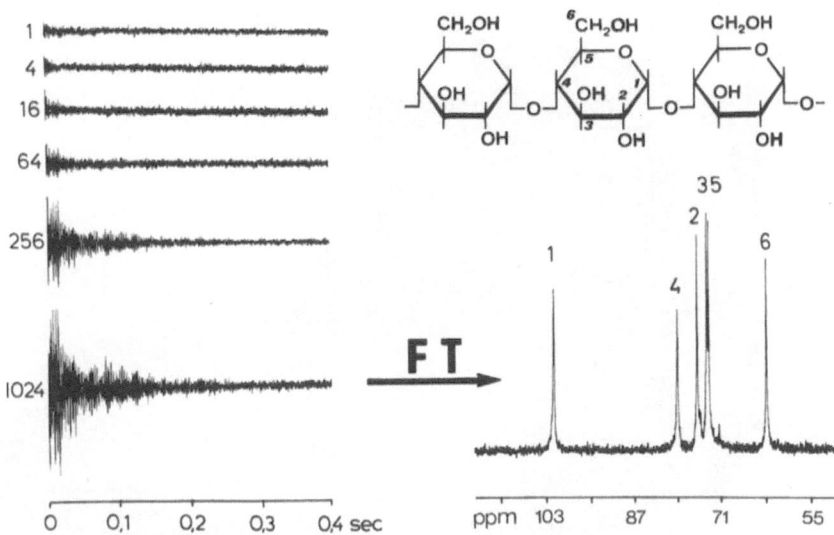

Fig. 27. Summation of NMR spectra. A small computer interfaced with the NMR system can carry out a Fourier transformation (*FT*) in a matter of seconds. Since this usually takes somewhat longer than the actual measurement of the decay curve, it is most efficient to sum the decay curves, thus improving the signal-to-noise ratio, and then perform a single Fourier transformation on the summed signal. In the present case the ^{13}C-NMR spectrum from an aqueous solution of amylose was analyzed. Amylose, like glycogen, is composed of a polymeric chain of glucose molecules, so that each carbon atom in the different glucose components produces a separate signal in the NMR spectrum

a process developed by the French mathematician J. B. J. Fourier in the early nineteenth century (Figs. 25, 26).

Because the small concentrations of substances in biologic material require a number of spectra added together in order to achieve a satisfactory signal-to-noise ratio, it is necessary first to sum the individual decay curves and then perform a single Fourier transformation on the summation decay curve (Fig. 27).

2.5 Design of a Whole-Body NMR System

Although a comprehensive discussion of the many physical and technological problems of a whole-body NMR system are beyond the scope of this introductory text, it is nevertheless important to review briefly the basic design principles of such systems and consider the function of their separate components.

The central component of a whole-body NMR system is the *magnet*, which has two main functions: First, it must generate a magnetic field of uniform intensity, i.e., the deviation from nominal field strength should not exceed 1 part in 10^{-5} over the region examined. Second, a useful aperture (bore) of at least 50 cm is required. The current-carrying coil that is used to generate the magnetic field may be a conventional electromagnet or a "superconducting" magnet.

In the conventional electromagnet, an array of four differently sized coils has

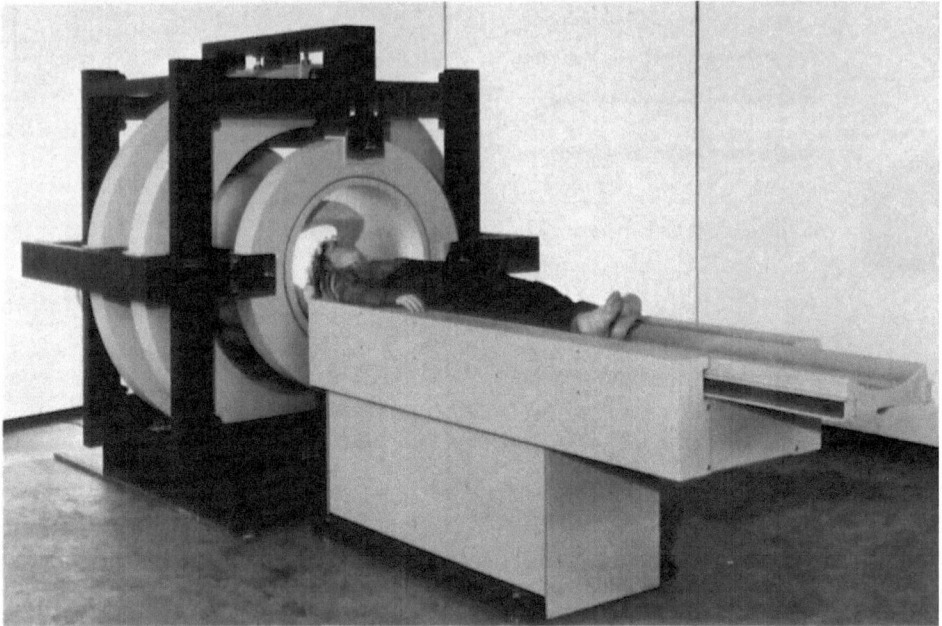

Fig. 28. Two pairs of different-size coils are used to generate a magnetic field of the highest possible uniformity. This system has a useful aperture of about 60 cm. (Photo from Bruker, Karlsruhe)

proved to be the best configuration for generating a magnetic field of optimum uniformity (Fig. 28).

To produce a field strength of 0.2 tesla, an electric current of 300 A (and about 200 V) must flow through the copper windings of the coil. Two technical problems are raised by this high degree of energy consumption: First, the power output needed to drive the magnet (about 60 kW) must be produced in a separate supply unit; second, a heavy-duty cooling circuit must be installed to dissipate the large amounts of heat generated in the magnet.

At present, technical constraints limit the power output of the conventional electromagnet to a field strength of about 0.25 tesla. When higher field strengths are desired, superconducting magnets must be used. The design of a superconducting magnet is similar to that of an ordinary electromagnet, except that the coil is constructed of a special niobium-titanium alloy. When cooled to the temperature of liquid helium (boiling point −269 °C or 4 K), this alloy becomes superconducting, meaning that its electrical resistance decreases to zero. Once a current has been established in the superconducting coil of the magnet, the current will continue to flow indefinitely, even after the power supply has been disconnected (Fig. 29).

Because the metal alloy abruptly loses its superconductivity at temperatures above 10K (−263 °C), the coil must be kept immersed in liquid helium. The temperature difference that exists between the interior of a superconducting magnet and the outside environment is approximately 300 °C. Consequently, the liquid helium would evaporate very quickly if precautionary measures were not taken. To reduce the consumption of liquid helium, the helium is surrounded by several evacuated

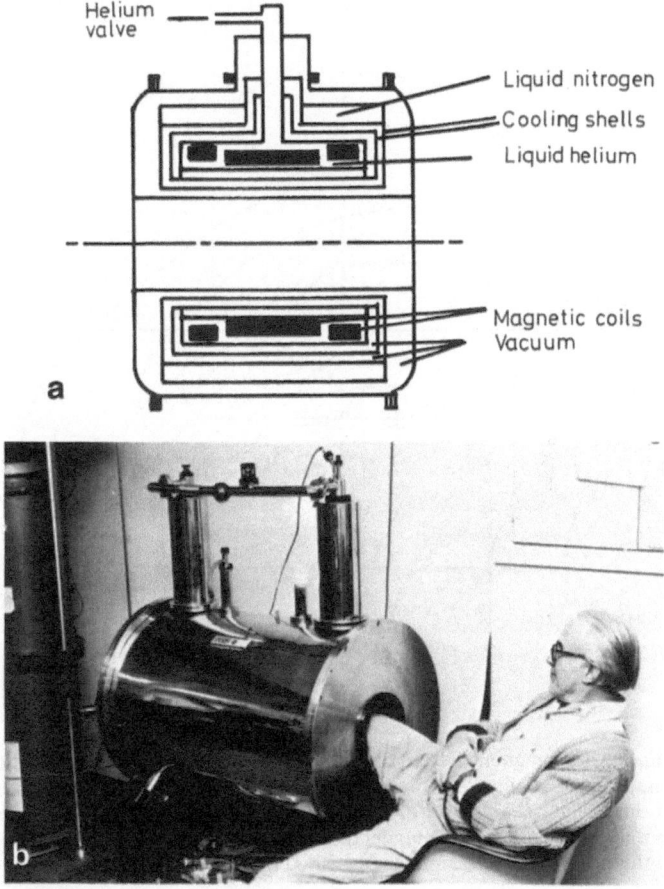

Helium
valve

Liquid nitrogen
Cooling shells
Liquid helium

Magnetic coils
Vacuum

a

b

Fig. 29 a, b. Design of a superconducting magnet (1.89 T). **a** The cutaway drawing of a horizontal-axis superconducting magnet shows the layered arrangement of its components. To reduce the rate of helium evaporation, the liquid helium is surrounded by several evacuated insulating layers and a helium- or nitrogen-cooled shell. **b** The photo shows Prof. B. Chance (Philadelphia) with his lower leg inserted into the 20 cm bore of a superconducting magnet, suitable for the NMR spectroscopy of laboratory animals and human limbs. (From documents furnished by Oxford Research Systems)

insulating layers (analogous to a vacuum bottle) as well as a layer of liquid nitrogen (boiling point $-196\,°C = 77$ K), which is less expensive than liquid helium.

Both magnet designs have their advantages and disadvantages, and it is not yet possible to make a definitive choice between the two. Superconducting magnets are far more expensive to construct than conventional electromagnets, and there are recurring costs for liquid helium and nitrogen. While electromagnets have a relatively low initial cost, the costs for electric current and cooling water can be substantial. Any future preference of one design over the other will depend largely on the evolution of these cost factors.

A major hazard to be considered with both magnet designs is a catastrophic breakdown of the magnetic field, as when an electromagnet suddenly loses its pow-

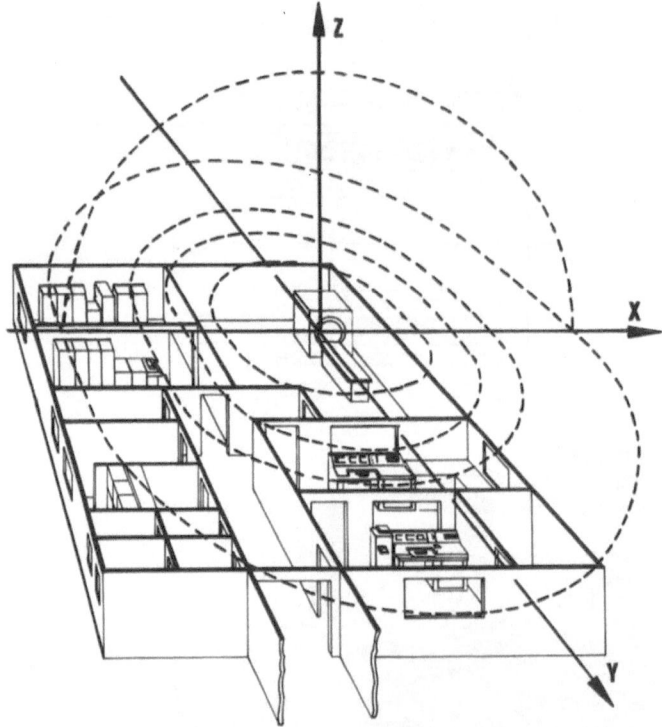

Fig. 30. The magnetic field from a whole-body magnet permeates the environment in all directions. The magnetic field strength at a distance of 7 m is approximately 1% the field strength at the center of the magnet. A field uniformity in excess of 1 part in 10^{-5} can be ensured only by maintaining a "safety zone" around the magnet that is free of moving ferrous objects. Adjacent floors are included in this zone. (From documents furnished by Siemens Co., Erlangen)

er supply or a superconducting magnet suffers an accidental "quench" (loss of superconductivity). In such cases the magnetic field strength falls to zero within microseconds, causing large and potentially harmful voltages and currents to be generated within the body of the patient. To prevent this, commercially available magnet systems come equipped with multiple safety devices which reduce the rate of magnetic field collapse to an acceptable level. Safety coils, for example, may be installed to induce an opposing magnetic field in the event of a main field breakdown. Large capacitors and parallel-connected diodes may also be used to slow the fall of magnetic field intensity.

In superconducting magnets, a magnetic field breakdown is additionally associated with a rapid evaporation of the liquid helium. Up to 100 m^3 of gaseous helium may be released from the system, greatly reducing the oxygen content of the air in the examining room. While helium vapors are not toxic or combustible, the room should nevertheless be vacated at once and thoroughly ventilated before it is reentered.

The magnetic field from a whole-body system is not confined to the region within the bore of the magnet. It extends for a considerable distance in all directions, its strength falling off gradually with distance from the magnet (Figs. 30, 31).

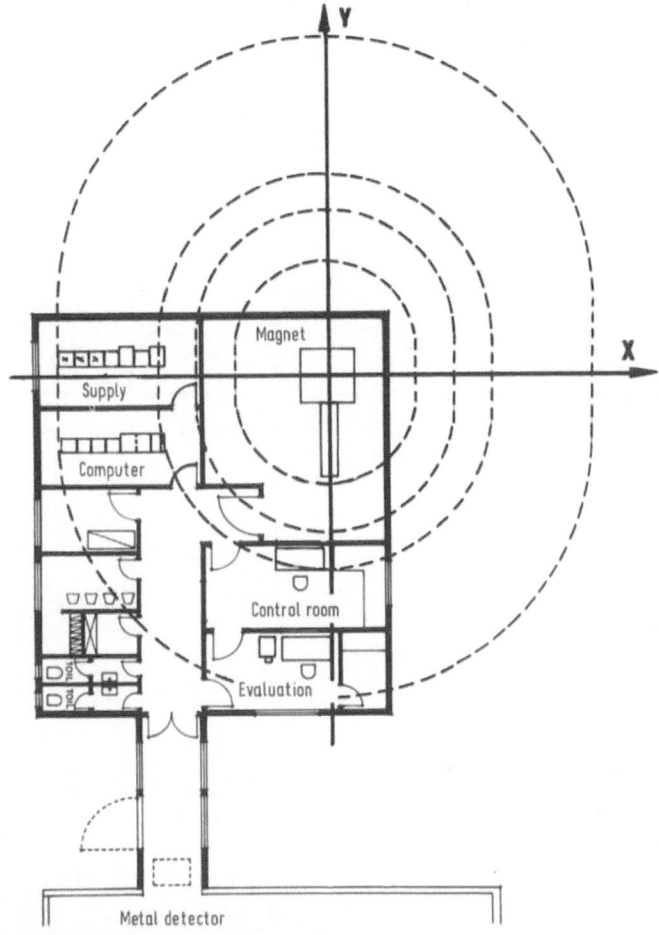

Fig. 31. Typical layout of a diagnostic NMR system. Much space is needed for the installation of most whole-body NMR systems. Components that are sensitive to magnetic fields, such as the computer and magnetic data storage media, must be kept a safe distance from the main magnet. With superconducting magnets, facilities must be available for the venting of helium vapors. (From documents furnished by Siemens Co., Erlangen)

At the same time, magnetic materials in the vicinity of the NMR system can distort the field pattern within the bore of the magnet and compromise the accuracy of the measurement. Therefore, special precautions must be taken with the environment of the system. The following recommendations are based upon a magnet with a field strength of 0.5 tesla at its center:

1. Movable iron-containing objects such as gas cylinders, ladders, guerneys, wheelchairs, etc. should be kept at least 5 m from the magnet in all directions (including adjacent floors).
2. Larger moving objects such as passing automobiles, elevators, etc. should be at least 15 m away.
3. Railroads and streetcars should be no closer than 30 m.

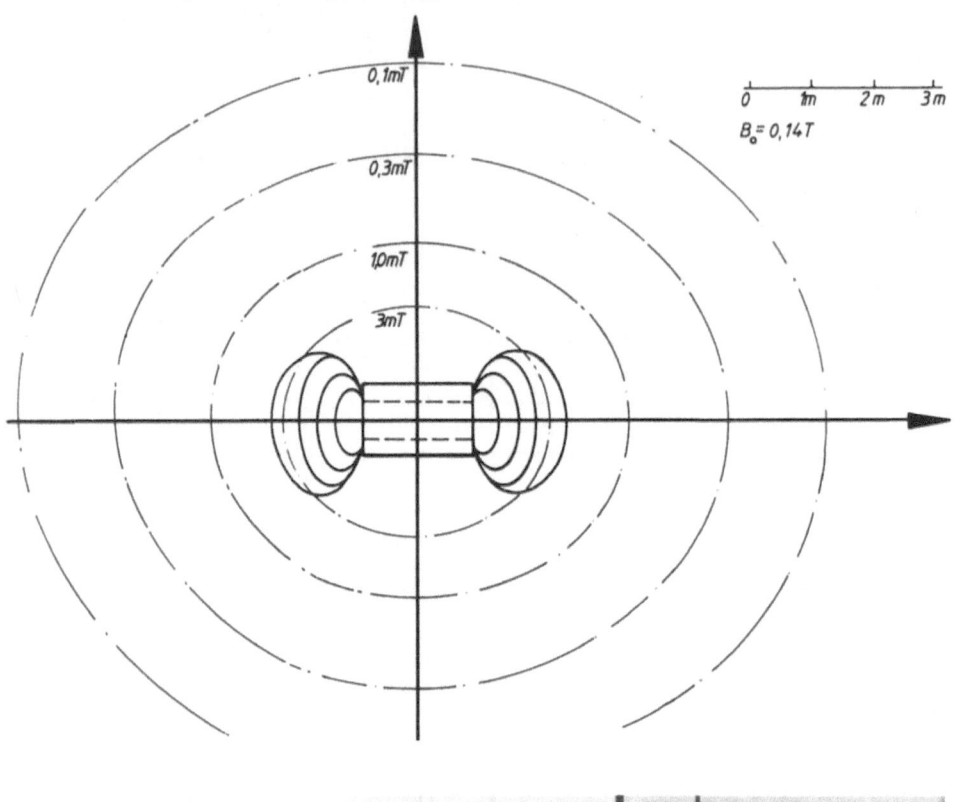

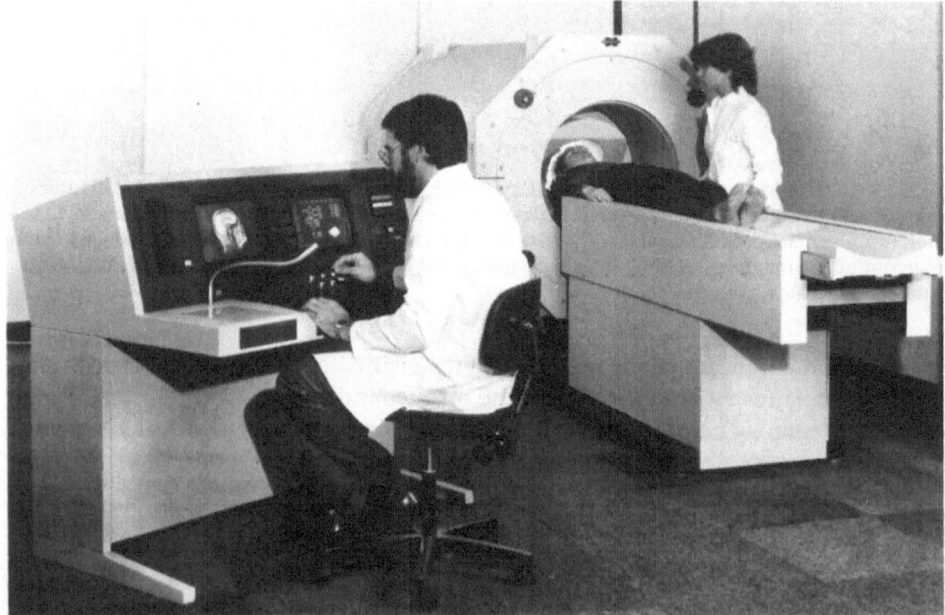

Fig. 32. A shielded whole-body magnet system. In this system an iron casing is used to redirect the magnetic field lines, thus greatly reducing the range and intensity of the stray field. Vertical to the main axis, the shielding effect is complete. (Photo from Bruker, Karlsruhe)

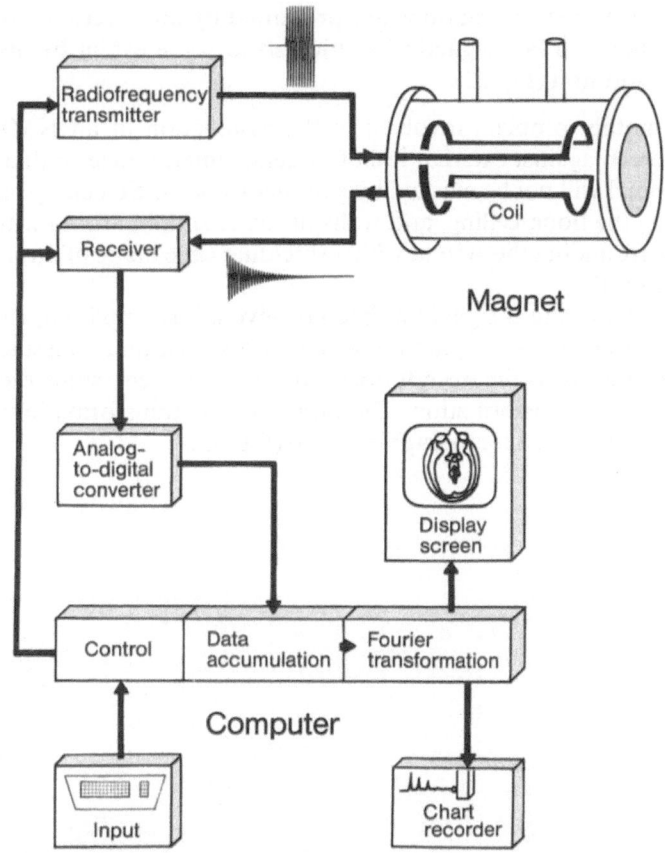

Fig. 33. Block diagram of an NMR system. A radiofrequency pulse is applied to the sample by the transmitter coil, which excites atomic transitions from the lower-energy to the higher-energy state. Reemission of the absorbed energy is picked up by the radiofrequency coil and measured by the receiver. The decay curve is digitized (analog-to-digital converter) and fed to the computer. After the individual scans are summed, a Fourier transformation is carried out, and the output is displayed as a spectrum or tomographic image

4. Sensitive electron-beam measuring instruments, photomultipliers, scintillation counters, oscilloscopes, monitors, etc. should not be used within 4 m of the magnet. Rheostat switches should not be used on lights.

5. Patients wearing cardiac pacemakers should come no closer to the magnet than 7 m.

6. Magnetic data-storage media (hard or floppy disks, magnetic tape) should be kept at least 5 m away.

7. The magnetic field may interfere with the operation of mechanical clocks and pocket calculators. Also, it may erase credit cards on which information is magnetically stored.

8. The environment of the magnet should be as free as possible of large, stationary ferrous objects. The effect of iron girders used in wall and ceiling construction can be largely offset by the installation of neutralizing coils ("shim coils").

9. If normal precautions are prevented by architectural constraints, the entire system can be magnetically shielded to some extent by installing several layers of soft iron (Fig. 32).

Another problem involved in the installation of an NMR system is the need to screen against external radiofrequency interference so that the low-intensity NMR signal will not be masked by extraneous noise. Screening may be done either by lining the floor, ceiling, and walls of the examining room with copper sheeting, or by surrounding the system with a shielding cage made of an rf-damping material (metal grid).

   Once the rf signal has been received, it is amplified, changed into binary digits by an analog-to-digital converter, and stored in a computer. The computer, which also controls the overall operation of the system, sums the signals and performs a Fourier transformation. The output of the transformation may be either an NMR spectrum or a tomographic image (Fig. 33).

3 NMR Spectroscopy in Intact Biologic Systems

The ^1H, ^{13}C and ^{31}P NMR spectra from a human forearm are compared in Fig. 34a. The ^1H spectrum contains only two signals, arising from tissue water and the CH_2 chains in fat. The ^{13}C spectrum covers a substantially broader range of chemical shifts and contains far more lines than the ^1H spectrum, but all signals may still be assigned to specific carbon atoms that are contained in tissue fat (Fig. 34b). In the ^{31}P spectrum, the signals are produced by ATP, phosphocreatine (PCr), and inorganic phosphate (P_i) (Fig. 34c).

In this chapter we shall examine and compare the spectra of various atomic nuclei in terms of their relevance to biological and medical investigations. It should be emphasized, however, that it has not yet been possible to perform whole-body spectroscopic measurements in human patients, because the superconducting magnets necessary for these investigations did not become available until the end of 1982. To date, NMR spectroscopic data are based entirely on studies of small laboratory animals and human limbs.

3.1 In Vivo ^1H-NMR Spectroscopy

The ^1H-NMR spectrum of intact tissues contains only two separate signals for tissue water and for the methylene hydrogen atoms in tissue fat. A quantitative assay of both components is easily performed and makes it possible to detect pathologic changes based upon the relative abundance of fat (Fig. 35).

Despite the high magnetic susceptibility of hydrogen nuclei and the relatively rapid rate at which hydrogen NMR spectra can be acquired, the clinical use of ^1H NMR spectroscopy is limited by the technical difficulties involved in the direct detection of metabolic products other than water and fat which are present in low concentrations (less than 10^{-5} mol/l). Hence, the major clinical application of hydrogen NMR is for diagnostic imaging, rather than spectroscopy.

3.2 In Vivo ^{13}C-NMR Spectroscopy

Because carbon-13 has a natural abundance of only 1%, ^{13}C-NMR spectra contain signals arising mainly from fat, although certain other compounds, such as choline and arginine in brain tissue, may also produce NMR signals (Fig. 36).

However, by using specifically labeled compounds in which the ^{13}C content has been increased from 1.1% to 90%, it is possible to detect a number of metabolic products occurring at lower concentrations. Because chemical reactivity is not affected by carbon-13, ^{13}C labeling is an excellent method for tracing the metabolic

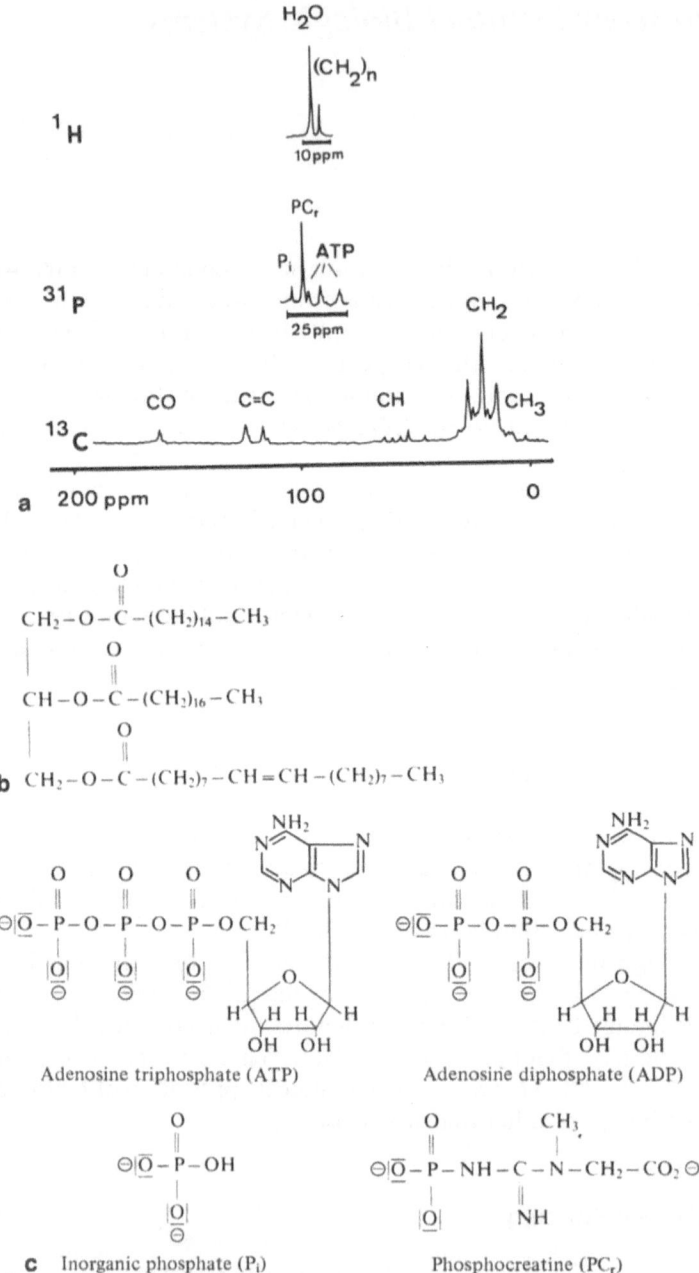

Fig. 34. a NMR spectra from the forearm of a live human subject. The peaks in the ^{1}H-NMR spectrum arise from the protons in water and in the methylene of tissue fat. The ^{31}P-NMR spectrum contains signals for inorganic phosphate *(Pᵢ)*, phosphocreatine *(PCr)* and the three different phosphorus atoms in adenosine triphosphate *(ATP)*. In the ^{13}C-NMR spectrum, all signals are produced by the various types of carbon atom occurring in fat (from documents furnished by Oxford Research Systems). **b** Glycerin (HOCH$_2$-CHOH-CH$_2$OH) esterified with palmitic, linoleic, and oleic acid *(top to bottom)*, a typical constituent of fat. **c** ATP, ADP, PCr, and Pᵢ

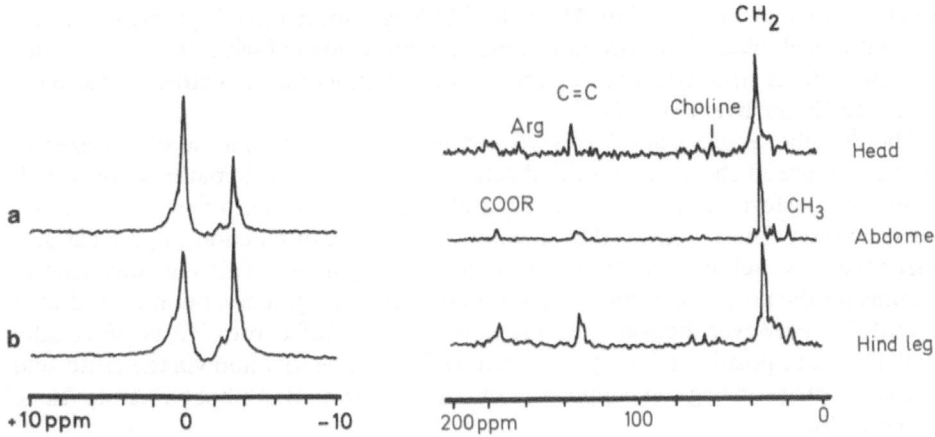

Fig.35a,b *(left)*. ^{1}H-NMR spectra of the human forearm. The ^{1}H spectrum of the forearm of a healthy 13-year-old boy (**a**) shows two signals arising from tissue water and fat. The intensity ratio of the two signals varies with the position of the radiofrequency coil, but the signal from water is always greater than the signal from fat. The ^{1}H-NMR spectrum from a 14-year-old boy with Duchenne muscular dystrophy (**b**) clearly demonstrates the pathologic increase in fat that is associated with this disease. Acquisition time for each spectrum was 8 s. [After Edwards et al. (1982) Lancet I: 725]

Fig.36 *(right)*. ^{13}C-NMR spectra from a live, anesthetized rat. In addition to the signals from fat, signals arising from choline and arginine can be detected in brain tissue. [After Alger JR et al. (1981) Science 214: 660]

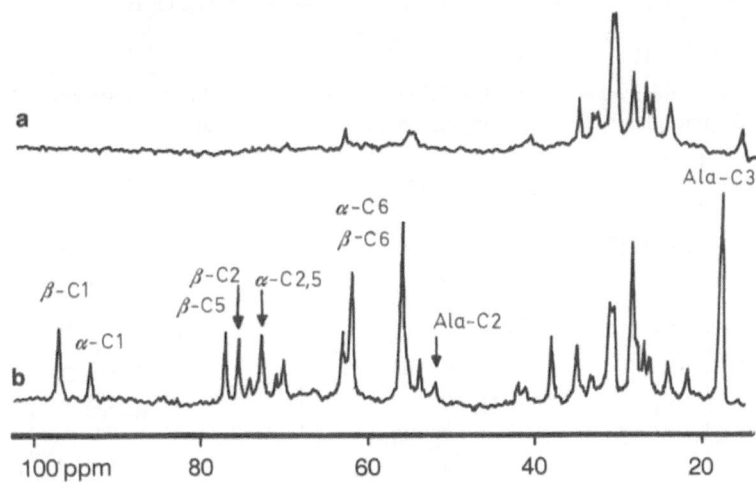

Fig.37a,b. ^{13}C-NMR spectra from an isolated, perfused mouse liver. **a** ^{13}C-NMR spectrum from a perfused mouse liver at 35 °C. **b** Two and one-half hours later another spectrum was obtained after two additions of 10 mmol (3–^{13}C) alanine and 20 mmol ethanol, 2 h apart. In addition to fat signals, this spectrum contains numerous signals from various metabolic products. For clarity, only the different carbon atoms of α and β glucose and of alanine have been labeled. [After Cohen SM et al. (1979) Proc Natl Acad Sci USA 76: 4808]

fate of a particular carbon atom. Thus, the ^{13}C NMR spectrum of a perfused mouse liver that was obtained after two additions of alanine labeled with ^{13}C at the C3 position provides a striking record of the various alanine-based syntheses that have taken place in the liver (Fig. 37).

The ^{13}C label appears at C1 and C6 and at C2 and C5 of glucose synthesized in the liver. Figure 38 shows a simplified scheme of the metabolic pathway by which alanine is transformed into glucose. First alanine is converted to pyruvic acid by transamination. This is followed by a complex sequence of reactions, known as gluconeogenesis, which culminates in the synthesis of glucose. This pathway readily accounts for the presence of the label at C1 and C6 of the glucose, but not for that at C2 and C5. However, because the measured content of carbon 13 is practically equal at all four positions, it may be concluded that isomerization via the citric acid cycle takes place during the conversion of oxaloacetic acid to phosphoenolpyruvic acid (Fig. 39).

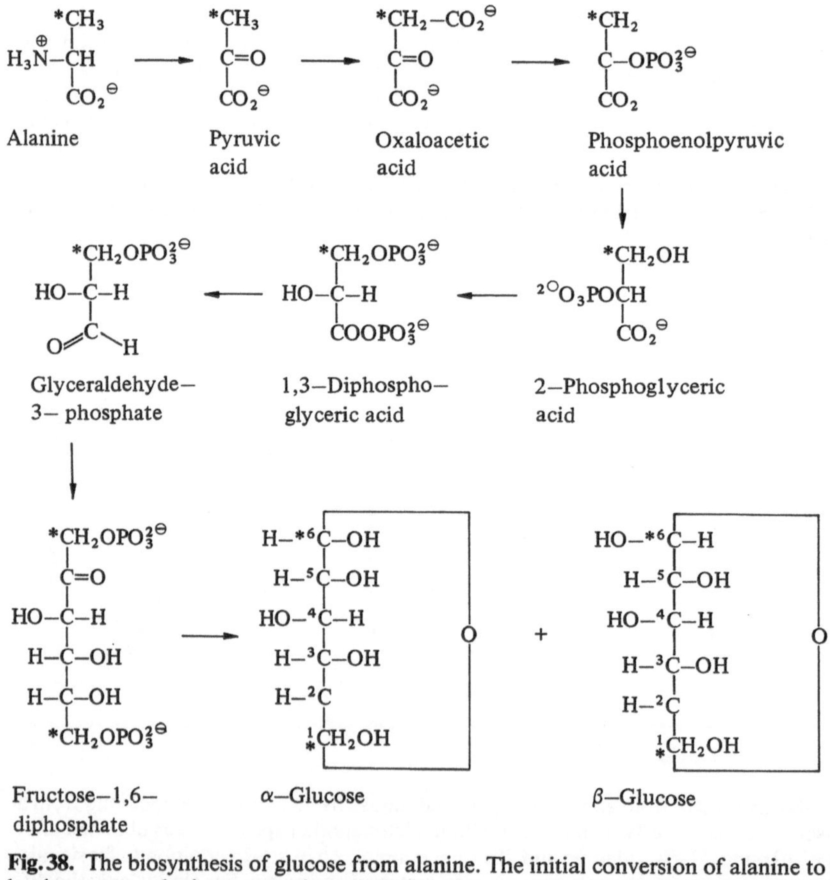

Fig. 38. The biosynthesis of glucose from alanine. The initial conversion of alanine to pyruvic acid by the enzyme alanine transaminase is followed by a sequence of reactions which ultimately yield glucose (gluconeogenesis). Starting from [3-^{13}C] alanine (each labeled carbon atom is marked with an *asterisk*), this multistep reaction pathway, shown here in simplified form, can account only for labeling at the 1 and 6 positions of glucose, which in aqueous solution is a mixture of the α and β forms

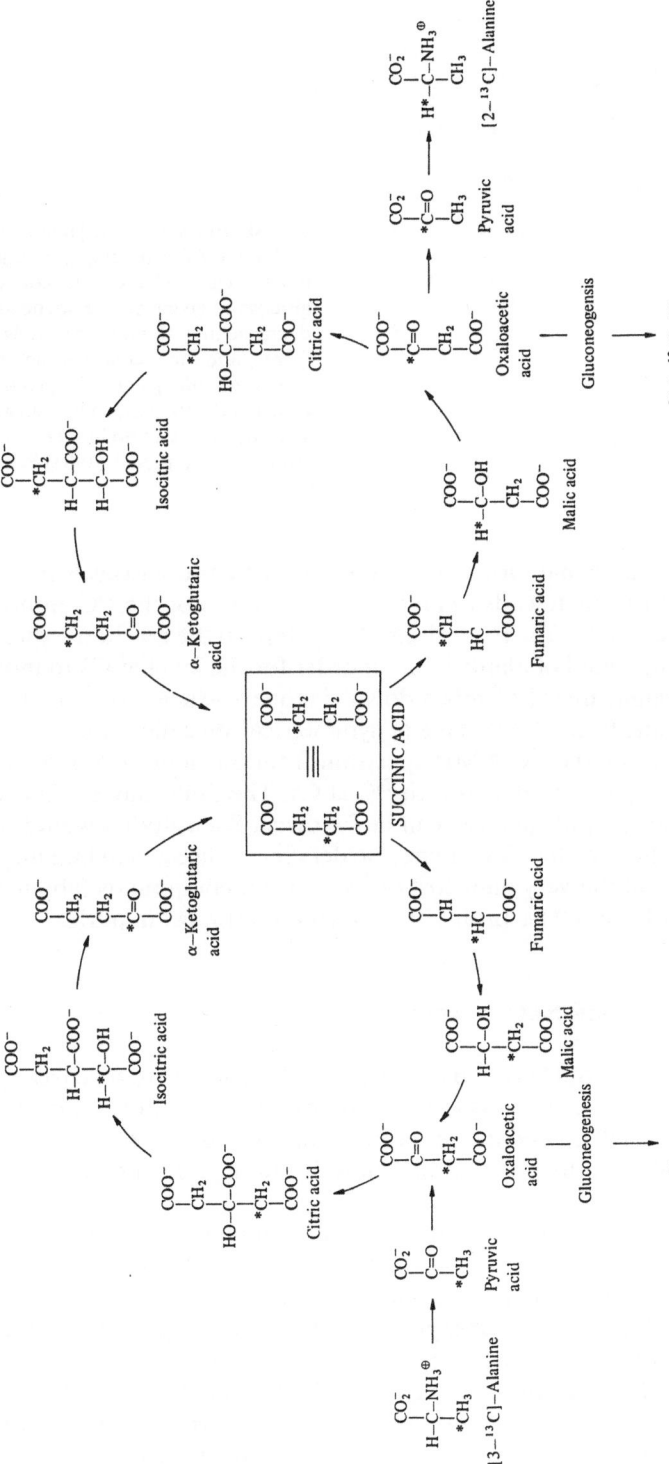

Fig. 39. The metabolic fate of labeled alanine C3. After conversion to pyruvic acid by transamination, the carbon framework of alanine enters the citric acid cycle at the stage of [3–^{13}C] oxaloacetic acid. Owing to the molecular symmetry of the succinic acid, the labeled carbon atoms become incorporated at C2 into the oxaloacetic acid, which is an intermediate both in the synthesis of glucose labeled at the 2.5 position and in the synthesis of [2–^{13}C] alanine

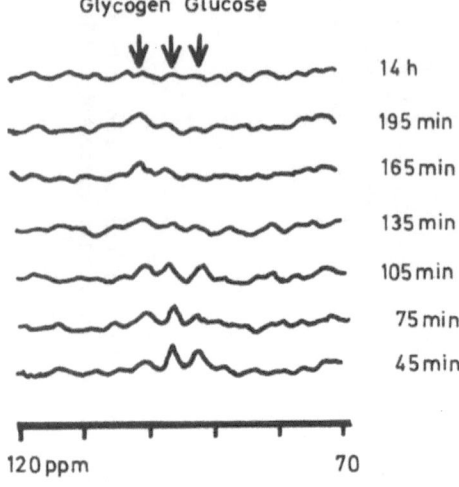

Glycogen Glucose

14 h

195 min

165 min

135 min

105 min

75 min

45 min

120 ppm   70

Fig. 40. In vivo ^{13}C-NMR spectra from a rat abdomen. After feeding the rat glucose labeled with ^{13}C at C1, the conversion of glucose to glycogen is evidenced by a decrease in both glucose signals (the α and β forms) and the increase in the liver glycogen signal. As time passes, the glycogen is broken down and enters the metabolism, and after 14 h the label can no longer be detected. [After Alger DR et al. (1981) Science 214: 660]

The succinic acid that forms in the course of the citric acid cycle possesses a molecular symmetry which leads to an equal distribution of the ^{13}C atoms to C2 and C3 in oxaloacetic acid. The reversibility of all steps in the reaction sequence is demonstrated by the signal of alanine C2. In order for the label at C3 to move to C2 of alanine, the alanine must be broken down via the citric acid cycle and then regenerated after passing through the stage of symmetrical succinic acid.

Figure 40 shows the ^{13}C NMR spectrum from the abdomen of an anesthetized rat after feeding glucose labeled with ^{13}C at C1. The time course of glucose storage as liver glycogen (= polyglucose) can be followed. Such studies would be of particular value for the investigation of enzyme deficiency diseases in human patients, although at present the very high cost of ^{13}C-labeled compounds (about US $ 500/g of glucose labeled at C1) is prohibitive for routine clinical inquiries.

3.3 In Vivo ^{31}P-NMR Spectroscopy

The phosphorus-31 nucleus is particularly useful for NMR spectroscopy applications, because the relative concentrations of the phosphorus metabolites ATP, phosphocreatine (PCr), and inorganic phosphate (P$_i$) define the energy status of the cell. A simplified outline of the biosynthesis and significance of the phosphorus compounds is shown in Fig. 41.

From the glycogen stored within the cell (a polyglucose), the enzyme glycogen phosphorylase splits off a terminal glucose molecule in the form of glucose-1-phosphate, thus initiating the process of glycolysis. The glycolytic pathway passes through several other sugar phosphates before yielding pyruvic acid and ATP, at which time NADH is formed from NAD$^+$. Following oxidative decarboxylation, the pyruvic acid is converted to acetyl-CoA. Then, in the citric acid cycle, the acetyl residue is broken down to carbon dioxide, and NADH and ATP are produced. In the subsequent respiratory chain, hydrogen is transferred from NADH to molecular

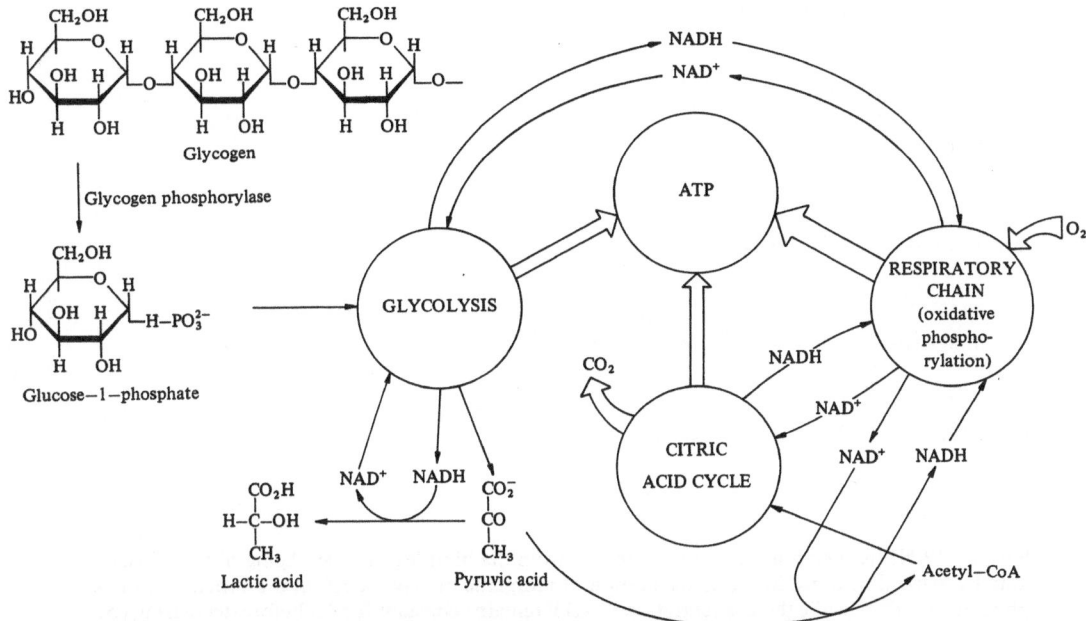

Fig. 41. Overview of the metabolic importance of phosphorus compounds. During glycolysis, stored glycogen is converted via glucose-1-phosphate to pyruvic acid, which then is broken down into carbon dioxide in the citric acid cycle, yielding energy that is chemically stored in the form of ATP. The active hydrogen in the NADH and FADH$_2$ molecules that form during this process is transferred to molecular oxygen in the respiratory chain, and the energy that is released is stored in 32 molecules of ATP. In the presence of an acute oxygen lack, adequate amounts of NAD$^+$ cannot be formed from NADH by the respiratory chain. NAD$^+$ is then produced by the conversion of pyruvic acid to lactic acid

oxygen, at which time NAD$^+$ is regenerated, while ATP is synthesized via oxidative phosphorylation. Overall, glycolysis generates only two molecules of ATP from one glucose molecule during glycolysis, the citric acid cycle yields two[1] molecules of ATP, and oxidative phosphorylation yields 32. Most of the chemical, osmotic, and mechanical work required at the various sites in the body is done by the conversion of ATP to ADP. ATP is the principal energy storehouse of the cell.

$$ATP \rightarrow ADP + P_i + 30.5 \text{ kJ/mol} \ (= 7.3 \text{ kcal/mol}).$$

However, because the ATP reserve in skeletal muscle is small and would be exhausted very rapidly during strenous exertion and ATP regeneration is relatively slow, ATP is stored in skeletal muscle indirectly in the form of phosphocreatine (PCr). Thus ATP can be synthesized very rapidly to meet the metabolic cellular requirements with the help of the enzyme creatine phosphokinase.

---

1 Actually, it is not ATP but GTP (guanosine triphosphate) which forms in the citric acid cycle. But the chemical energy of GTP is readily transferred to ATP with the help of nucleoside diphosphokinase: GTP + ADP → GDP + ATP

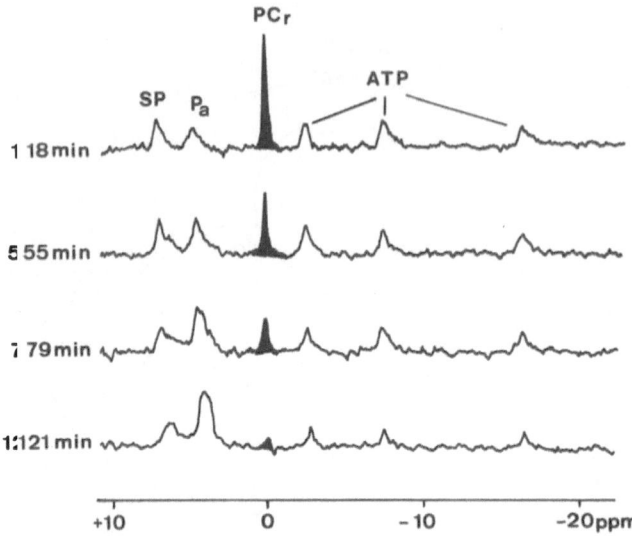

Fig. 42. ^{31}P-NMR spectrum of an intact muscle from the hind leg of a rat. Aging of the muscle is initially marked by a rise in the concentration of inorganic phosphate *(P$_i$)* at the expense of phosphocreatine *(PCr)*, while the concentration of ATP remains constant for 2 h before decreasing (*SP*, sugar phosphates). Because the position of the phosphocreatine resonance is pH-independent in the physiologic range and is demonstrable in many intact tissues, the chemical shifts of all signals are referred to PCr as an arbitrary zero. [After Hoult DI et al. (1974) Nature 252: 285]

$$PCr + ADP \xrightarrow{\text{PCr-Kinase}} Creatine + ATP$$

Figure 42 shows the ^{31}P-NMR spectrum of an intact muscle freshly dissected from the hind leg of a rat. A quantitative analysis of sugar phosphates, inorganic phosphate, phosphocreatine and ATP is easily performed by measuring the areas under the resonance peaks. As the isolated muscle tissue ages, the concentration of inorganic phosphate increases at the expense of phosphocreatine, while the ATP concentration is initially kept constant by creatine phosphokinase. After about 2 h, when all the phosphocreatine has been used up, the ATP concentration declines.

Thus, the endogenous metabolism of biopsied tissue continuously changes the concentration of phosphorus metabolites, reflecting various in vivo enzyme activities and providing information on a number of pathogenic changes. Figure 43 shows the results of a ^{31}P-NMR study of muscle samples taken from a healthy subject and from a patient with neuromuscular disease.

In each case the time course of the concentration changes in the phosphorus metabolites shows the general features observed for all isolated muscle: a fall of phosphocreatine, a rise of inorganic phosphate, and a fall of ATP after the consumption of phosphocreatine. Initially the cell compensates for the ATP consumption by the breakdown of phosphocreatine, while glycolysis is accelerated in an effort to synthesize new ATP, causing the concentration of sugar phosphates to rise. As the tissue ages, however, the enzyme systems that support glycolysis gradually fail, and the resulting failure in the breakdown of sugar phosphates causes their

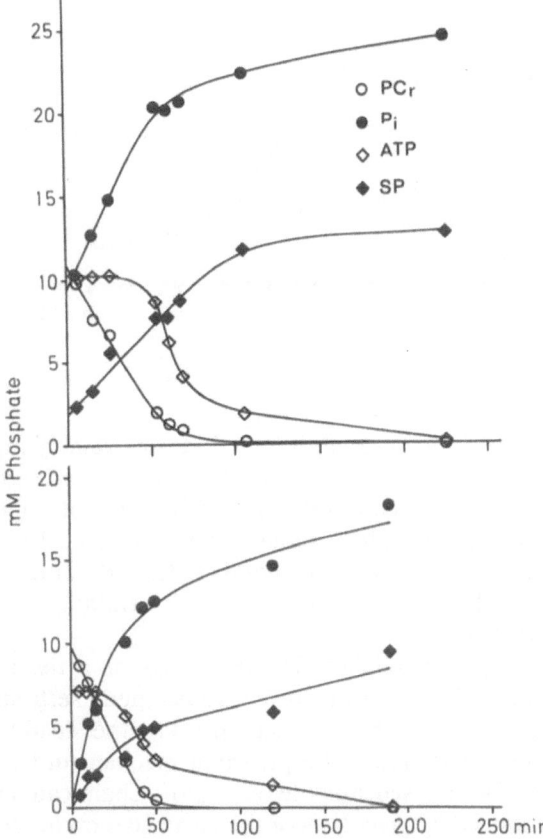

Fig. 43. ^{31}P-NMR spectra from human muscle samples. These spectra were used to follow the time course of the concentrations of the various phosphorus compounds in normal muscle *(top)* and in muscle taken from a patient with neuromuscular disease *(bottom)* (*PCr*, phosphocreatine; P_i, inorganic phosphate; *SP*, sugar phosphates; *ATP*, adenosine triphosphate). The total phosphorus content of the diseased muscle is reduced by half compared to the normal tissue. The much more rapid fall of ATP reflects the reduced working capacity of the diseased muscle. [After Burk CT et al. (1977) Science 195: 145]

concentration to remain constant. In the diseased muscle, the NMR spectrum shows that the total phosphorus content is initially reduced by half compared with normal muscle. In addition, the phosphocreatine in the diseased muscle is broken down much more rapidly to inorganic phosphate via ATP, so that the period during which the ATP level can be held constant by the breakdown of phosphocreatine decreases from about 50 min in normal muscle to 24 min in diseased muscle. Thus, the ^{31}P-NMR spectrum reflects the markedly reduced working capacity of the pathologically altered muscle tissue.

The noninvasive measurement of intracellular pH is a powerful tool for the study of biochemical processes. Because pH has a major influence on the rate of many enzyme-catalyzed reactions, the hydrogen ion may be regarded as a key metabolite. The dependence of the chemical shift of the inorganic phosphate peak on hydrogen ion concentration forms the basis of pH measurement by ^{31}P NMR spectroscopy (Fig. 44).

The pK of the second dissociation step of phosphoric acid lies within the physiologic pH range of 6–7.5.

$$H_2PO_4^- \rightleftharpoons HPO_4^{2-} + H^+ \quad pK = 6.8$$

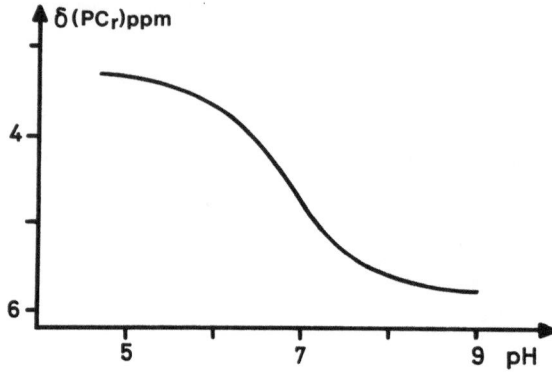

Fig. 44. pH Dependence of the chemical shift of the P_i resonance. With increasing pH, the chemical shift of the P_i signal in the ^{31}P-NMR spectrum decreases, the pH dependence showing the typical behavior of a titration curve. The chemical shift is referred to the PCr signal as zero

Because the dissociation reaction is extremely rapid, $H_2PO_4^-$ and HPO_4^{2-} do not produce two separate signals but an averaged signal, whose chemical shift depends on the relative concentrations of the two molecular species. Thus, the position of the inorganic phosphate signal shows the typical behavior of a titration curve whose point of inflection corresponds to the pK value. The function shown in Fig. 44 may be used as a calibration curve for determining the intracellular pH to an accuracy of 0.05–0.01.

The intracellular pH can change as a result of sudden muscular exertion in which ATP is consumed. With sustained exertion, even the supply of phosphocreatine would not be adequate to offset the steady loss of ATP. Normally, oxidative phosphorylation is the principal mechanism by which ATP is synthesized. However, during ischemia the respiratory chain can no longer function, and ATP is only produced by glycolysis or the citric acid cycle; both pathways require the presence of NAD^+ as an oxidant, and an intact respiratory chain is necessary for NAD^+ production. In order for ATP synthesis to be maintained by glycolysis, pyruvic acid has to be reduced to lactic acid. In this process, NADH is oxidized to NAD^+, which is then utilized for glycolysis. The intracellular pH decreases due to lactic acid accumulation.

Figure 45 illustrates how the ^{31}P NMR measurement of intracellular pH can be applied to clinical diagnostics. A patient with suspected McArdle's syndrome (bottom) and a normal control subject (top) each performed arm exercises (flexion of fingers at 2-s intervals) while ^{31}P-NMR spectra were continuously recorded from their forearms. After 2 min exercise under ischemic conditions (inflated sphygmomanometer cuff on the upper arm), the pH in the muscle tissue of the control subject decreased by 0.6, while the phosphocreatine content showed a marginal increase. Under the same experimental conditions, the patient's resting spectrum showed a slightly elevated intracellular pH at rest, and this remained virtually unchanged during ischemic exercise. The phosphocreatine, meanwhile, showed a profound decrease after only 1 min of exercise.

The failure of lactic acid production in the skeletal muscle of patients with McArdle's syndrome is due to a lack of glycogen phosphorylase (see Fig. 41). As a result of this enzyme deficit, glycolysis cannot function adequately during exercise

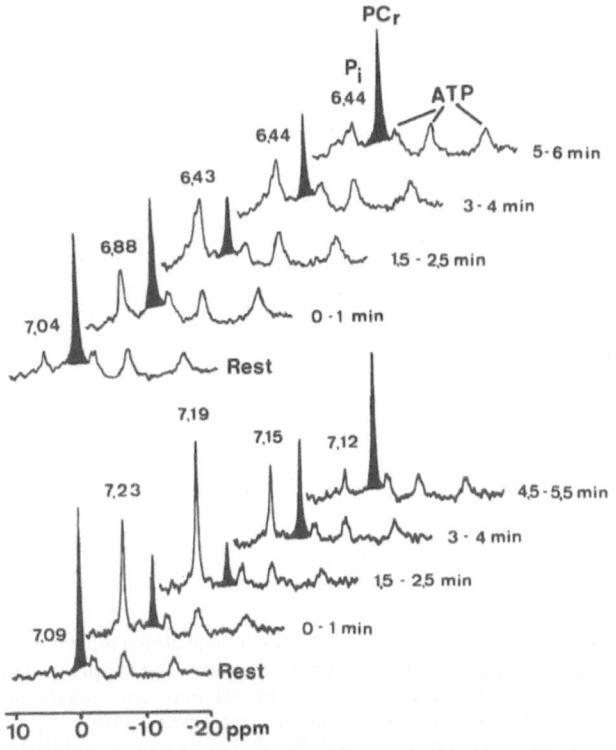

Fig. 45. ^{31}P-NMR spectra from the forearm of a healthy subject *(top)* and from a patient with McArdle's syndrome *(bottom)*. First the arm muscles were exercised for 1 min, during which time ischemia was produced by a tourniquet on the upper arm. Then the arm was rested, and 3 min later arterial flow was restored. The spectra are displayed in a staggered fashion for clarity. Intracellular pH was determined from the precise resonance frequency of the inorganic phosphate signal *(P$_i$)*. [After Ross BD et al. (1981) N Engl J Med 304: 1338]

and cannot compensate for ATP consumption. Without the conversion of pyruvic acid to lactic acid and NAD$^+$, the intramuscular pH remains unchanged.

Ischemia is generally associated with a fall of the intracellular pH. If an infarction is induced by ligating the LAD of a perfused rabbit heart, the fall of pH from 7.2 to 6.4 leads to a complete failure of glycolysis, as the essential enzyme systems are no longer operative at this range of pH. If the oxygen supply is then restored by releasing the ligature (Fig. 46), the pH quickly returns to the original, normal value owing to the removal of lactic acid, while the phosphocreatine and ATP levels, which reflect the energy state of the tissue, return to normal at a considerably slower rate.

The rate of this normalization process can be followed by measuring the change in pH and in the ATP and PCr levels, and it depends on many physiologic parameters. This raises a question that is of great importance to medicine: How can organs be protected against irreversible damage from a transient oxygen deficit like that associated with surgical procedures and organ transplantations? ^{31}P NMR spectroscopy can possibly supply some answers to this question, as it gives direct informa-

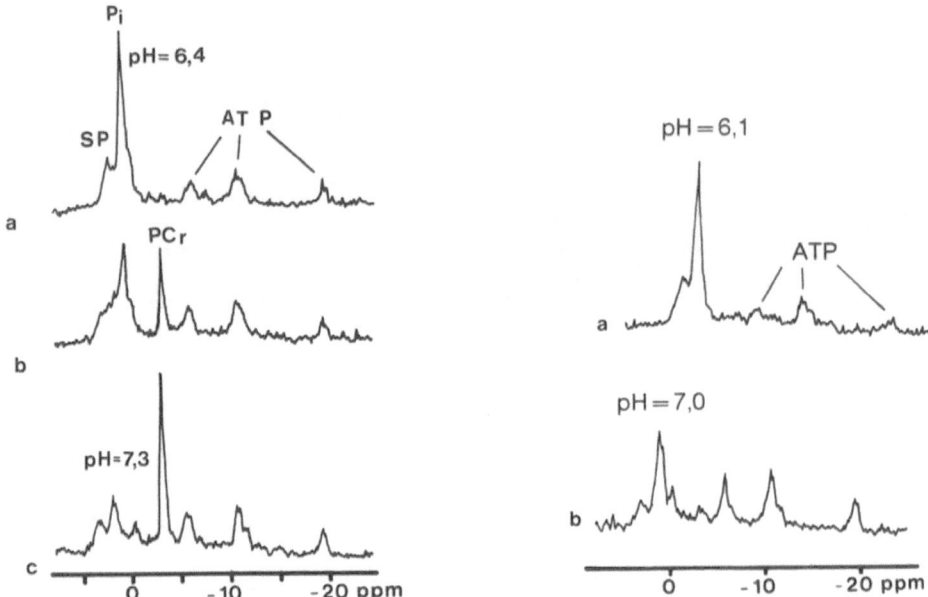

Fig. 46 a–c *(left).* Recovery of a rabbit heart from global infarction. **a** 31P-NMR spectrum of the perfused rabbit heart after 40 min global infarction. The supply of high-energy phosphocreatine is exhausted, and ATP is also decreased. The pH value calculated from the exact position of the inorganic phosphate resonance indicates the overacidification of the myocardial tissue. **b** Two minutes after the oxygen supply is restored, the pH has returned to normal owing to lactic acid removal, and the synthesis of phosphocreatine from inorganic phosphate has commenced. **c** Six minutes after restoration of the cardiac oxygen supply the phosphocreatine level has normalized, while the ATP concentration has not yet reached its normal level. [After Hollis DP (1979) Bull Magn Reson 1: 27]

Fig. 47 a, b *(right).* 31P-NMR spectra of a rabbit heart with global infarction. **a** After 40 min global infarction, the spectrum indicates a complete failure of energy metabolism. Phosphocreatine is absent, and only traces of ATP are apparent. The pH has fallen from 7.2 to 6.1. **b** In this heart potassium chloride solution was administered as a cardioplegic agent before infarction was induced. After 40 min the energy metabolism is excellent compared with that of the untreated heart. [After Hollis DP (1978) J Magn Reson 29: 319]

tion about the physiologic state of an organ. In one experiment, the 31P-NMR spectrum of a perfused rabbit heart was recorded after 40 min of complete ischemia (Fig. 47). The spectrum showed no signal from phosphocreatine and only a weak signal from ATP; the pH had fallen to 6.1. By injecting potassium chloride solution as a cardioplegic agent, the fall of pH and ATP could be slowed to a dramatic degree. As Fig. 47 shows, the energy metabolism of the cardiac tissue after 40 min global infarction was excellent compared with the untreated heart.

To simulate a renal transplantation, Sehr et al. excised a rat kidney and stored it in ice-cold buffer (Fig. 48). Even at the low temperature, the ATP content fell rapidly under ischemic conditions. Meanwhile, the pH of the kidney tissue changed very little owing to the slowed rate of glycolysis. After the kidney was connected to the blood circuit of a second rat, it was possible to follow the time course of the recovery of ATP synthesis in the reperfused organ. Because the ability of a transplanted

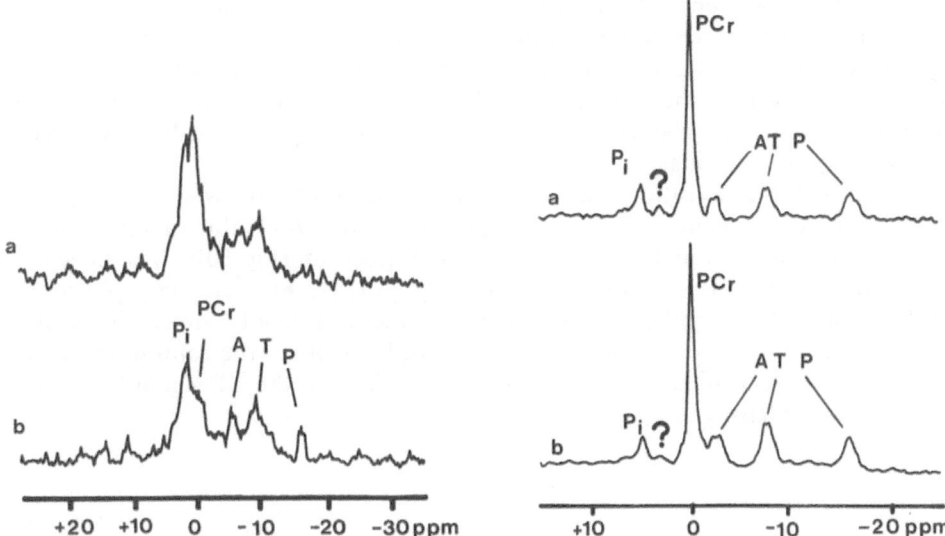

Fig. 48 a, b *(left).* ^{31}P-NMR spectra of a rat kidney for which transplantation was simulated. After 15 min ischemia in cold buffer, the freshly excised rat kidney was connected to the circulation of a second rat. The regeneration of the energy metabolism, i. e., the resynthesis of ATP from inorganic phosphate, can be monitored on the basis of the ^{31}P-NMR spectra. Spectrum **a** was started together with perfusion. Spectrum **b** was initiated after 51 min perfusion. [After Sehr PA et al. (1977) Biochem Biophys Res Commun 77: 195]

Fig. 49 a, b *(right).* ^{31}P-NMR spectra from the forearm muscles of a hypothyroid patient. Spectra were recorded with 300 pulses at 2-s intervals. **a** Besides the signals for ATP, PCr and P_i, the spectrum contains an unidentified signal in the phosphodiester range. **b** After 2 months' treatment with L-thyroxine the intensity of this signal is notably decreased, correlating well with a general regression of hypothyroid symptoms. [After Iles RA et al. (1982) Progr NMR spectros 15: 49]

organ to resynthesize high-energy metabolites is an important measure of its viability, experiments of this kind are invaluable for exploring new ways of prolonging the preservation of organs intended for transplantation.

Besides the major phosphorus metabolites ATP, P_i, and PCr, diseased tissues may contain abnormally large amounts of other phosphorus-containing compounds as a result of metabolic abnormalities. In the ^{31}P-NMR spectrum from the forearm of a hypothyroid patient (Fig. 49), we find that the usual signals for ATP, P_i, and PCr are accompanied by an additional resonance in the phosphodiester region of the spectrum. The exact structure of the metabolite responsible for this signal is not known. But whatever its nature, the intensity of the signal provides a convenient and objective index for monitoring the patient's response to subsequent therapy. After 2 months' treatment with L-thyroxine, the intensity of the signal decreased markedly. This correlated well with a general improvement in the patient's clinical condition.

By employing an extension of ^{31}P NMR spectroscopy, it is possible to measure the rates of certain enzyme-catalyzed reactions. In this technique, called "saturation transfer," a selected peak in the spectrum is saturated by applying the corresponding resonance frequency. When an NMR experiment is subsequently performed, the equal energy-level distribution of the selected nuclei leads to an extinction of the signal (see Fig. 19). If the saturated atom is now chemically transformed by means of a chemical reaction, the saturation will be transferred to the NMR signal, which corresponds to the new chemical shift. Thus, saturation transfer may be viewed as a form of magnetic labeling.

Figure 50 illustrates saturation transfer in perfused rat heart. By applying energy at the resonance frequency of the γ-phosphorus atom in ATP, this signal becomes magnetically saturated and disappears from the spectrum (Fig. 50b). The intensity of the phosphocreatine signal is simultaneously reduced. Because the γ-phosphorus atom of ATP is rapidly transferred to creatine and thus becomes chemically transformed into the phosphorus atom of phosphocreatine, the saturation of the γ-^{31}P signal of ATP is transferred to the PCr signal, and the ^{31}P signal from PCr has a reduced intensity.

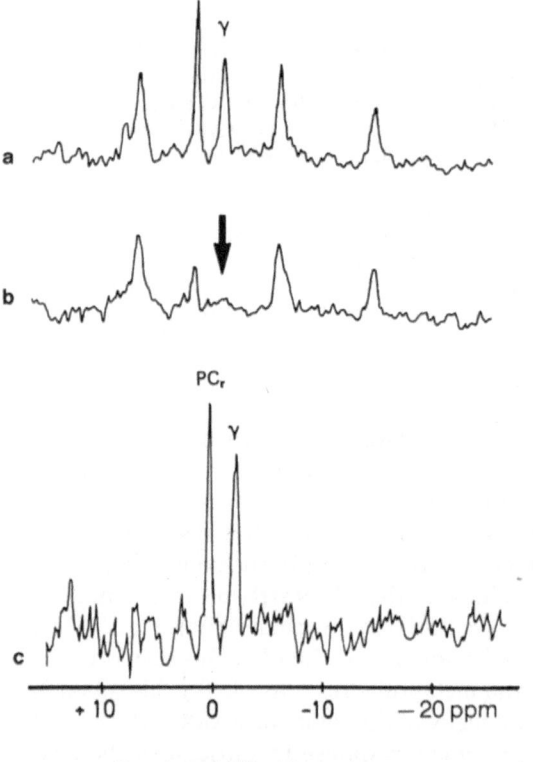

Reaction of ATP with creatine

Adenosine–O–PO$_2$–O–PO$_2$–OPO$_3$ + Creatine →

    [ATP]   + [Cr]

Adenosine–OPO$_2$$^{\ominus}$–O–PO$_3$$^{2\oplus}$ + Creatine–O–PO$_3$$^{2\ominus}$

    [ADP]   + [KP]

d

Fig. 50 a–d. ^{31}P-NMR saturation transfer in an isolated, perfused rabbit heart. By selective irradiation with radiofrequency energy, the γ-phosphorus atom of ATP is completely saturated. In a subsequent NMR experiment, this leads to an extinction of the signal (**b**). Because the γ-phosphorus atom of ATP undergoes a constant exchange with the phosphorus atom of phosphocreatine *(PCr)*, the PCr signal is also weakened. This is most evident in the difference spectrum (**c**), which was obtained by the subtraction of spectra **a** and **b**. By quantitative evaluation, the rate of the chemical reaction can be estimated. **d** The reaction of ATP with creatine

Thus, while the measurement of concentration on the basis of signal intensity makes it possible to determine the *equilibrium constant,* the *rate* of the forward and reverse reactions can be estimated by an accurate evaluation of the saturation transfer experiment.

The rate of the reverse reaction $PCr + ADP \rightarrow ATP + Cr$ can be measured in the same manner. In the perfused rat heart, the following reaction rates can be measured under physiologic conditions:

$$PCr + ADP \rightarrow ATP + Cr \qquad 4.0 \text{ mmol/s}$$
$$ATP + Cr \quad \rightarrow ADP + PCr \qquad 2.9 \text{ mmol/s}$$
$$ATP \qquad \rightarrow ADP + P_i \qquad 1.1 \text{ mmol/s}$$

Laboratory studies of dissected muscle or perfused organs can yield much information on the metabolism of healthy and diseased tissue under various physiologic conditions, but they cannot replace the noninvasive, selective NMR spectroscopic examination of individual organs in situ.

In a conventional NMR spectrometer, the object to be analyzed is completely surrounded by the radiofrequency coil (Fig. 51), and the recorded spectrum represents the sum of all resonating nuclei in the volume enclosed by the coil (Fig. 51 a). This method is not suitable for the selective investigation of individual organs in situ. However, by mounting special "field profiling" coils within the bore of the mag-

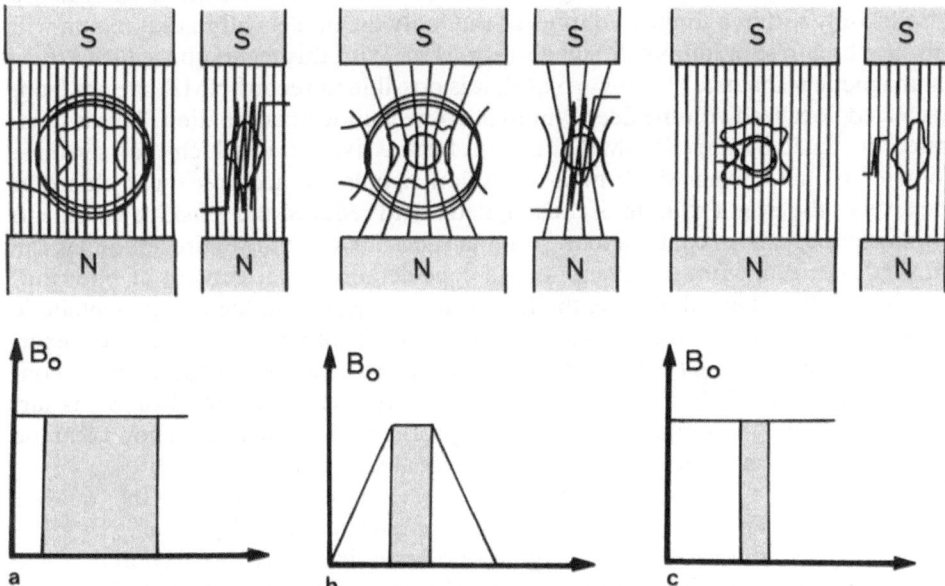

Fig. 51 a–c. Schematic diagrams of NMR systems for the in vivo examination of biologic materials. **a** Conventional NMR spectrometer. The observed signal intensity corresponds to the total concentration in the volume enclosed by the radiofrequency coil. **b** Topical NMR spectrometer. Profiling coils are used to shape the magnetic field so that only the nuclei within a circumscribed volume produce high-resolution signals. **c** Surface coils. By using a suitable coil geometry and energy distribution around the coil, NMR spectra can be recorded from a localized region

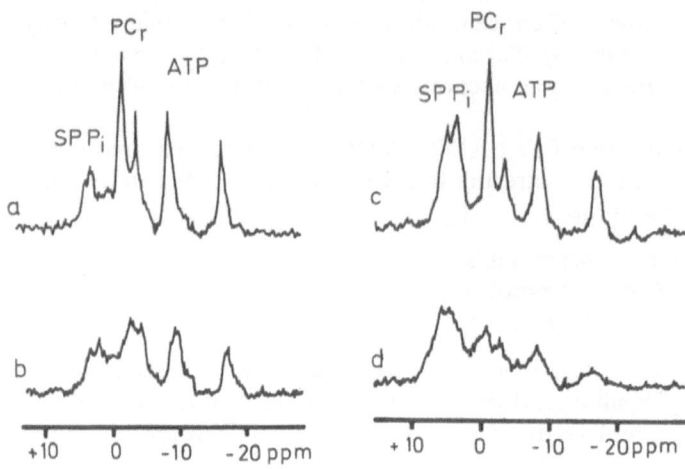

Fig. 52 a–d. ^{31}P-NMR spectra from a live rat. During recording of the spectra, the magnetic field was profiled so that the sensitive volume was localized to the region of the liver. **a** Normal spectrum. **b** Spectrum obtained after field profiling. Targeting on the organ of interest can be confirmed from the PCr signal, since the liver contains very little phosphocreatine. **c** Normal spectrum after surgical interruption of the hepatic blood supply. **d** Same as **c**, but with field profiling. [After Gordon RE et al. (1980) Nature 287: 736]

net, it is possible to shape the magnetic field in such a way that the field lines are parallel only within a limited volume, so that only the nuclei within that region will produce high-resolution NMR signals (Fig. 51 b). With this technique, called "topical magnetic resonance" (*topos* = place), it is possible to record NMR spectra from a selected, localized volume deep within the object under examination. This is illustrated in Fig. 52 by the ^{31}P-NMR spectrum from a live rat, in which the "sensitive volume" (volume from which high-resolution signals are acquired) was placed at the level of the liver. Profiling of the magnetic field reduced the sensitive volume to a diameter of 2 cm. In contrast to the normal spectrum from the entire abdomen, the profiled spectrum from the liver shows a considerably lower content of phosphocreatine. Unlike skeletal muscle, the liver is not subject to sudden, large metabolic demands, so it does not require a special supply of rapidly available chemical energy in the form of PCr. ATP production via normal pathways is adequate for the constant metabolic work that the liver must perform. If the hepatic blood supply is surgically interrupted to simulate ischemia, the profiled NMR spectrum shows that the increase in inorganic phosphate occurs entirely at the expense of ATP.

Another way of obtaining spectra from selected body regions is by the "surface coil" technique. The surface coil, whose working principle is shown schematically in Fig. 51 c, does not surround the region to be examined, but is placed against the body surface and permits a selective study of tissues adjacent to the coil. For example, the ^{31}P-NMR spectrum of the brain can be recorded in vivo by positioning the coil on the surface of the head (Fig. 53).

This example is of major importance in view of the extreme difficulty of performing in vivo brain studies by other methods. The in vivo value of 1.7 measured for the PCr:ATP ratio on the basis of ^{31}P-NMR spectra is higher than the values

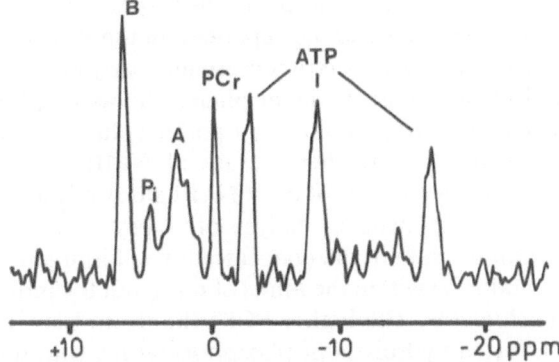

Fig. 53. ^{31}P-NMR spectrum of the brain of a 17-day-old infant. The spectrum was acquired with a surface coil (5-cm diameter) with profiling of the main magnetic field. The sensitive volume was 4 cm in diameter, its center being 2 cm above the plane of the coil (total volume of about 30 cm^3), and mainly encompassed the parietal and temporal lobes. A total of 1024 pulses were applied at intervals of 2.26 s (total acquisition time 38 min). The coil was tuned to 32.5 MHz. Besides the signals from *ATP*, inorganic phosphate *(P$_i$)* and phosphocreatine *(PCr)*, there are two signals, labeled A and B, which probably are assignable to glycerol-3-phosphorylcholine and glycerol-3-phosphoryl-ethanolamine *(A)*, and to ribose-5-phosphate *(B)*. From the position of the P$_i$ signal, an intracellular pH of $6.72 \pm$ SD 0.11 may be inferred. The PCr:P$_i$ ratio, reflecting the state of the energy metabolism, was equal to 1.7. [After Wilkie DR et al. (1983) Lancet I: 1059]

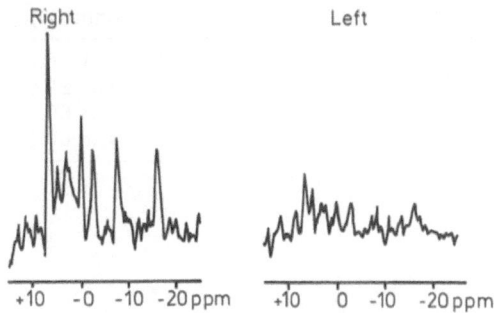

Fig. 54. ^{31}P-NMR spectra of the brain of a 22-day-old infant. As in Fig. 53, a 5-cm surface coil and profiled field were used. Spectra were recorded with 128 pulses at 10.26-s intervals. Marked differences are apparent in the spectra from the two cerebral hemispheres. The total phosphorus content of the left hemisphere is only 40% that of the right. This decrease reflects a loss of brain tissue that was confirmed when large cysts were found on ultrasound examination at 4 weeks of age. [After Wilkie DR et al. (1983) Lancet I: 1059]

previously obtained by rapid freezing and subsequent laboratory analysis. This suggests that even a careful dissection may entail some degree of irreversible tissue trauma. Surface coil ^{31}P spectra have also shown that the concentration of inorganic phosphate in intact brain tissue is substantially lower (1.5 mmol/l) than was previously assumed (4–5 mmol/l).

During the routine ultrasound examination of a 14-day-old infant with no clinical abnormalities, a diffuse echodensity was detected in the left cerebral hemisphere. The ^{31}P-NMR spectra obtained at 22 days (Fig. 54) showed that the total phosphorus content in the left hemisphere was only 40% of that in the right. They also showed marked discrepancies in the PCr : P_i ratio (left 0.8, right 1.4). It was assumed that these spectra represented early evidence of a significant degeneration of brain tissue in the left hemisphere. This assumption was confirmed by the sonographic demonstration of a large porencephalic cyst at 2 weeks later.

To illustrate further the use of NMR for quantitative investigations, Table 3 compares the results of a ^{31}P NMR study of the human forearm in vivo with the results of a biochemical analysis of biopsy material. Both methods yield identical values for the sum of the concentrations of inorganic phosphate and phosphocreatine (P_i and PCr) within the limits of error, but the individual values for P_i and PCr differ considerably. The higher PCr value from the ^{31}P-NMR study proves that biopsy leads to a hydrolysis of phosphocreatine, with the formation of creatine and inorganic phosphate.

Finally, I shall relate an example of how ^{31}P NMR spectroscopy can be used to assess the efficacy of pharmacologic treatment. Figure 55 shows the experimental arrangement used to investigate a perfused heart by the surface-coil technique. A lo-

Table 3. Quantitative assay of phosphorus metabolites by ^{31}P NMR spectroscopy compared to conventional postbiopsy analysis. (After Edwards et al. [69])

	^{31}P NMR (mmol/l)	Biopsy (mmol/l)
ATP	5.1 ± 0.1	5.5 ± 0.07
PCr	28.5 ± 0.4	17.4 ± 0.2
P_i	4.3 ± 0.2	10
PCr + P_i	32.8 ± 0.4	31.6 ± 3.3

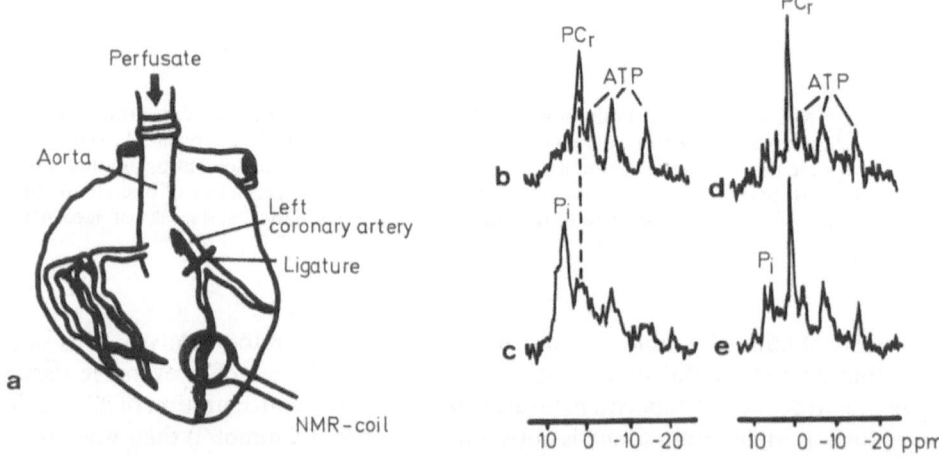

Fig. 55 a–e. ^{31}P-NMR spectroscopy of a perfused heart using a surface coil. **a** Diagram of the local infarction. **b** Spectrum from a normal perfused rat heart. **c** Spectrum after local infarction. **d** Spectrum from a normal heart pretreated with verapamil. **e** Same as **d**, but after local infarction. [After Nunnally RL and Bottomley PA (1981) Science 211: 177]

cal infarction was simulated by ligating the left anterior descending coronary artery. The ^{31}P-NMR spectrum obtained 40 min after ligation demonstrates an almost complete failure of energy metabolism in the cardiac tissue. In hearts that were pretreated with the calcium antagonist verapamil before the ligature was applied, the ^{31}P-NMR spectrum obtained under the same conditions shows that the energy metabolism was excellent compared with that in the untreated hearts after equal periods of infarction.

In summary, ^{31}P NMR spectroscopy provides an accurate method for determining the concentrations of the high-energy phosphorus within intact cells. So far, technical constraints have to limited experiments to laboratory animals, human limbs, and infants. However, the variety of studies that have been done to date, ranging from noninvasive pH measurements and the detection of ischemic tissue damage to the diagnosis of enzyme deficiency states and the assessment of organs for transplantation, underscore the great potential of ^{31}P NMR spectroscopy in the field of human medicine. The necessary technical groundwork for in vivo investigations, such as the use of profiled fields and surface coils, has already been laid, so clinical applications will depend essentially on the ability of industry to develop powerful whole-body magnets with high field strengths. Given the substantial development and production costs of these systems, however, it appears that only a *combination* of whole-body NMR spectrometry and NMR tomography will have a realistic prospect for routine clinical use. Developmental work by industry in this area is already underway.

References for Chap. 1–3

Abraham A (1961) The principles of nuclear magnetism. Clarendon, Oxford (A standard work written mainly for physicists)
Becker ED (1980) High resolution NMR. Theory and chemical applications, 2nd edn. Academic Press, New York
Günther H (1980) NMR Spectroscopy. Wiley, Chichester.
Shaw D (1976) Fourier transform NMR spectroscopy. Elsevier/North Holland, Amsterdam
Slichter CP (1978) Principles of magnetic resonance. Springer, Berlin Heidelberg New York

Textbooks on Biochemical Applications of NMR Spectroscopy

Dwek RA (1973) Nuclear magnetic resonance in biochemistry. Oxford University Press, London
Govil G, Hosur RV (1982) NMR: Basic principles and progress, vol 20: Conformation of biological molecules: New results from NMR. Springer, Berlin Heidelberg New York
James TL (1975) Nuclear magnetic resonance in biochemistry: Principles and applications. Academic Press, New York
Jardetzky O, Roberts GCK (1981) NMR in molecular biology. Academic Press, New York
Knowles PF, Marsh D, Rattle HWE (1976) Magnetic resonance of biomolecules. Wiley, London
Opella SJ, Lu P (1979) NMR and biochemistry. Dekker, New York Basel
Wüthrich K (1976) NMR in biological research: Peptides and proteins. North Holland, Amsterdam

Monographs and Progress Reports

NMR in Biology. Dwek RA, Campbell ID, Richards RE, Williams RJP (eds) Academic Press, London, 1977
Biological applications of magnetic resonance. Shulman RG Academic Press, New York, 1979

NMR of intact biological systems. Williams RJP, Andrew ER, Radda GK (eds) Philos Trans R Soc Lond [Biol] 289: 379–559 (1980)
Magnetic resonance in biology. Cohen JS. Wiley & Sons, New York, 1980
NMR in medicine. Damadian RR (Hrsg) Springer, Berlin Heidelberg New York, 1981
Nuclear magnetic resonance and its applications to living systems. Gadian DG, Oxford University Press, Oxford, 1982
Noninvasive probes of tissue metabolism. Cohen JS (ed) Wiley & Sons, New York, 1982

Review Articles on in Vivo NMR Spectroscopy

Shulman RG et al. (1979) Cellular applications of ^{31}P and ^{13}C nuclear magnetic resonance. Science 205: 160
Burt CT, Cohen SM, Barany M (1979) Analysis of intact tissue with ^{31}P NMR. Annu Rev Biophys Bioeng 8: 1
Hollis DP (1980) Phosphorus NMR of cells, tissues and organelles. Biol Magn Reson 2: 1
O'Neill IK, Richards CP (1980) Biological phosphorus-31 NMR spectroscopy. Annu Rep NMR Spectr 10 A: 133
Gadian DG, Radda GK (1981) NMR studies of tissue metabolism. Annu Rev Biochem 50: 69
Shaw D (1981) In vivo chemistry with NMR. In: Kaufman L, Crooks LE, Margulis AR (eds) Nucl. Magn. Reson. Imaging Med. Igaku-Shoin, Tokyo, pp 147–183
Roberts JKM, Jardetzky O (1981) Monitoring of cellular metabolism by NMR. Biochim Biophys Acta 639: 53
Iles RA, Stevens AN (1982) NMR studies of metabolites in living tissue. Prog Nucl Magn Reson Spectrosc 15: 49
Gronenborn A, Roth K (1982) NMR-Spektroskopie in vivo. Chem i. U. Zeit 16: 1
Gordon RE, Hanley PE, Shaw D (1982) Topical magnetic resonance. Prog Nucl Magn Reson Spectrosc 15: 1
Bradbury EM, Radda GK (1983) NMR techniques in medicine. Ann Intern Med 98: 514
Shulman RG (1983) NMR spectroscopy of living cells. Sci Am 248: 86
Shaw D (1983) In vivo topical magnetic resonance. Org Magn Reson 21: 225

Selected Original Publications

Heart

Studies of acidosis in the ischemic heart by phosphorus NMR. Garlick P, Radda GK, Seeley JP, Biochem J 184: 547 (1979)
NMR studies of cancer and heart disease. Hollis DP, Bull Magn Reson 1: 27 (1979)
NMR of phosphorus in the perfused heart. Hollis DP, IEEE Trans Nucl Sci 27: 1250 (1980)
Phosphorus-31 NMR studies of the energetic state and intracellular pH of the isolated rat heart during ischemia. Rossi A et al., J Physiol (Paris) 76: 902 (1980)
Studies of metabolism in the isolated, perfused rat heart using carbon-13 NMR. Bailey IA et al., FEBS Lett 123: 315 (1981)
The effects of reperfusion on the phosphorus-31 NMR spectrum of ischemic rat hearts. Bailey IA et al., Biochem Soc Trans 9: 234 (1981)
A phosphorus-31 NMR study of the effects of reflow on the ischemic rat heart. Bailey IA et al., Biochim Biophys Acta 637: 1 (1981)
A phosphorus-31 NMR study of metabolism in the hypoxic perfused rat heart. Matthews PM, Biochem Soc Trans 9: 236 (1981)
The steady-state rate of ATP synthesis in the perfused rat heart measured by phosphorus-31 NMR saturation transfer. Matthews PM et al., Biochem Biophys Res Commun 103: 1052 (1981)
Measurement of free magnesium in perfused and ischemic arrested heart muscle. A quantitative phosphorus-31 NMR and multiequilibrium analysis. Wu ST et al., Biochemistry 20: 7399 (1981)
Gated Sodium-23 NMR images of an isolated perfused working rat heart. DeLayre IL et al., Science 212: 935 (1981)

A phosphorus-31 NMR study of the metabolic and functional effects of changes in extracellular calcium on the isolated perfused rat heart. Seymour AM et al., Biochem Soc Trans 9: 475 (1981)

Assessment of pharmacological treatment of myocardial infarction by phosphorus-31 with surface coils. Nunnally RL, Bottomley PA, Science 211: 177 (1981)

A phosphorus-31 NMR saturation transfer study of the regulation of creatine kinase in the rat heart. Matthews PM, Biochim Biophys Acta 721: 312 (1982)

Effects of l- and d-propranolol on the ischemic myocardial metabolism of the isolated guinea pig heart, as studied by phosphorus-31 NMR. Nakazawa M et al., J Cardiovasc Pharmacol 4: 700 (1982)

The effects of insulin on myocardial metabolism and acidosis in normoxia and ischemia. A phosphorus-31 NMR study. Bailey IA, Biochim Biophys Acta 720: 17 (1982)

Phosphorus NMR spectroscopy of cardiac and skeletal muscles. Ingwall JS, Am J Physiol 242: H 729 (1982)

In vivo carbon-13 NMR studies of heart metabolism. Neurohr KJ et al., Proc Natl Acad Sci USA 80: 1603 (1983)

Kidneys

Study of rat kidney in vivo during hypovolemic shock by ^{31}P-NMR. Chan L et al., Biochem Soc Trans 9: 239 (1981)

Phosphorus NMR study of the rat kidney in vivo. Balaban RS et al., Kidney Int 20: 575 (1981)

The role of intrarenal pH in regulation of ammoniagenesis: ^{31}P-NMR studies of the isolated perfused rat kidney. Ackerman JJH et al., J Physiol (Lond) 319: 65 (1981)

Phosphorus-31 NMR analysis of the renal response to respiratory acidosis. Freeman D et al., Biochem Soc Trans 10: 399 (1982)

Energetics of sodium transport in the kidney. Saturation transfer phosphorus-31 NMR. Freeman D et al., Biochim Biophys Acta 762: 325 (1983)

Liver

Phosphorus-31 NMR studies on membrane phospholipids in microsomes, rat liver slices and intact perfused rat liver. Kruijff BDe et al., Biochim Biophys Acta 600: 343 (1980)

Effects of fructose on the energy metabolism and acid-base status of the perfused starved-rat liver: A ^{31}P-NMR study. Iles RA et al., Biochem J 192: 191 (1980)

Direct proton and natural abundance carbon-13 NMR observation of liver changes induced by ethionine. Brock RE, Biochem Biophys Res Commun 108: 940 (1982)

Metabolic interrelationships of intracellular pH measured by double-barrelled microelectrodes in perfused rat liver. Cohen RD et al., J Physiol (Lond) 330: 69 (1982)

Hepatic metabolism by phosphorus-31 NMR. Iles RA, Griffiths JR, Biosci Rep 2: 735 (1982)

Structure and metabolism of mammalian liver glycogen monitored by carbon-13 nuclear magnetic resonance. Sillerud LO, Shulman RG, Biochemistry 22: 1087 (1983)

Muscle Tissue

Studies of the biochemistry of contracting and relaxing muscle by the use of ^{31}P-NMR in conjunction with other techniques. Dawson MJ et al., Philos Trans R Soc Lond [Biol] 289: 445 (1980)

Natural abundance carbon-13 NMR spectra of intact muscle. Doyle DD et al., FEBS Lett 131: 147 (1981)

High-resolution proton magnetic resonance spectra of muscle. Yoshizaki K et al., Biochim Biophys Acta 678: 283 (1981)

Phosphorus-31 NMR studies of energy metabolism and tissue pH in ischemic rat leg. Thulborn KR, Biochem Soc Trans 9: 237 (1981)

Mitochondrial regulation of phosphocreatin/inorganic phosphate ratio in exercising human muscle: A gated phosphorus-31 NMR study. Chance B et al., Proc Natl Acad Sci USA 78: 6714 (1981)

NMR analysis of intact tissue including several examples of normal and diseased human muscle deter-

minations. Glonek T et al., NMR in Medicine, Damadian R (ed) Springer, Berlin Heidelberg New York, 1980

Simultaneous in vivo measurement of oxygen utilization and high-energy phosphate metabolism in rabbit skeletal muscle by Multinuclear proton and phosphorus-31 NMR. Thulborn KR et al., J Magn Reson 45: 362 (1981)

Generation of phosphodiesters during fast-to-slow muscle transformation. A phosphorus-31 NMR study. Burt CT et al., Biochim Biophys Acta 721: 492 (1982)

Phosphorus-31 NMR of contractile systems. Barany M, Glonek T, Methods Enzymol 85: 624 (1982)

Application of phosphorus-31 NMR spectroscopy to the study of striated muscle metabolism. Meyer RA et al., Am J Physiol 242: C1 (1982)

Phosphorus-31 NMR studies of control of mitochondrial function in phosphofructokinase-deficient human skeletal muscle. Chance B et al., Proc Natl Acad Sci USA 79: 7714 (1982)

Preliminary observations on the metabolic responses to exercise in humans, using phosphorus-31 NMR. Ross BD et al., Ciba Found Symp 87: 145 (1982)

Nuclear magnetic resonance studies of forearm muscle in Duchenne dystrophy. Newman RJ et al., Br Med J 284: 1072 (1982)

Clinical use of NMR in the investigation of myopathy. Edwards RHT et al., Lancet I: 725 (1982)

Quantitation of lactid acid in caffeine-contracted and resting frog muscle by high resolution natural abundance carbon-13 NMR. Doyle DD, Barany M, FEBS Lett 140: 237 (1982)

Brain

Localized noninvasive detection and description of ischemic cerebral damage using NMR. Fossel ET, Ingwall JS, Cerebrovasc Dis 12: 91 (1981)

Cerebral energy metabolism in rats studied by phosphorus nuclear magnetic resonance using surface coils. Bottomley PA et al., Magn Reson Imaging 1: 81 (1982)

Phosphorus-31 NMR saturation transfer measurements of the steady state rates of creatin kinase and ATP synthease in the rat brain. Shoubridge EA et al., FEBS Lett 140: 288 (1982)

Developmental changes of creatine kinase metabolism in rat brain. Nowood WI et al., Am J Physiol 244: C 205 (1983)

In vivo phosphorus-31 NMR studies on experimental cerebral infarction. Naruse S et al., Jpn J Physiol 33: 19 (1983)

Non-invasive investigation of cerebral metabolism in newborn infants by phosphorus nuclear magnetic resonance spectroscopy. Cady EB et al., Lancet I (8333): 1059 (1983)

Neoplastic Tissue

Use of the NMR of nuclei other than proton in tumor studies. Granger P, J Biophys Med Nucl 5: 137 (1981)

Phosphorus-31 NMR investigation of solid tumors in the living rat. Griffiths JR et al., Biosci Rep 1: 319 (1981)

Human tumors as examined by in vivo phosphorus-31 NMR in athymic mice. Evanochko WT et al., Biochem Biophys Res Commun 109: 1346 (1982)

Phosphorus-31 NMR spectroscopy of in vivo tumors. Ng TC et al., J Magn Reson 49: 271 (1982)

NMR studies of tumors. Griffiths JR, Iles RA, Biosci Rep 2: 719 (1982)

4 NMR Tomography

The development of a tomographic technique based on the nuclear magnetic resonance phenomenon appears the most interesting and promising innovation of recent years in medical imaging. Even the first industrially produced prototype instruments presented cross-sectional images of the human body of outstanding quality when compared with first-generation X-ray computed tomograms. The excellent anatomic resolution without the need for ionizing radiation is the reason why NMR tomography[1] has attracted such great interest among medical professionals. Nevertheless, NMR imaging is still in the developmental stage, and for the time being optimism should be guarded. The suitability of NMR tomography as a routine diagnostic method has yet to be established by clinical trials that have only recently begun.

A particular difficulty in the interpretation of NMR tomograms lies in the complex correlation between measured image intensities and NMR-relevant tissue properties such as water content, relaxation times, and flow velocity (in vessels). Thus, the optimum interpretation of NMR tomograms and the proper selection of imaging parameters require an understanding of the physical principles of the technique. The present chapter is concerned not only with the practical capabilities of NMR tomography in the clinical setting, but also with the basic physical principles which underlie the technique of NMR imaging.

4.1 Basic Principles of NMR Imaging

To date, all commercially developed NMR tomographic systems produce images on the basis of NMR signals arising from tissue water. This is due both to the high NMR sensitivity of the hydrogen atom and to the well-known abundance of water in human tissues. Thus, the discussion begins with a look at an NMR spectrum which contains only one signal, that of tissue water.

In a conventional NMR experiment, the object to be examined is placed within a highly homogeneous magnetic field, i. e., one in which the field strength is equal at all points within the volume of interest. In this case all hydrogen atoms in the tissue water will have the same resonance frequency, regardless of their spatial position

---

1 A number of terms have been applied to the creation of an image with NMR: NMR imaging, spin imaging, spin mapping, NMR zeugmatography, NMR tomography. Because these terms vary somewhat in their meanings and are not standardized, shall use "NMR tomography" to refer to all imaging techniques

within the object. The relationship between the resonance frequency v_0 and the magnetic field strength B_0 is given by the Lamor relation (see Section 2.1)

$$v_0 = \frac{\gamma}{2\pi} B_0 \qquad\qquad (9\,a)$$

where γ is the magnetogyric ratio of the nucleus.

First let us consider the NMR experiment as applied to a tube filled with water (Fig. 56). By irradiating the tube with radiofrequency energy at the resonance frequency of the water, *all* hydrogen atoms in the region of interest, shown in Fig. 56 as an idealized plane, will be on resonance, and the resultant spectrum will contain a single, sharp water signal. In order to form an image, it is necessary to provide spatial resolution. This can be done by superimposing on the main field B_0 a magnetic field "gradient" (G_x, G_y, or G_z) which varies in a linear manner along a particular axis. When this is done, the hydrogen atoms at different positions will experience different magnetic field strengths, and, according to the Larmor relation, they will have different resonance frequencies. The net result of this is a broadening of the water signal. The signal intensity at a particular position v_0 in the spectrum is pro-

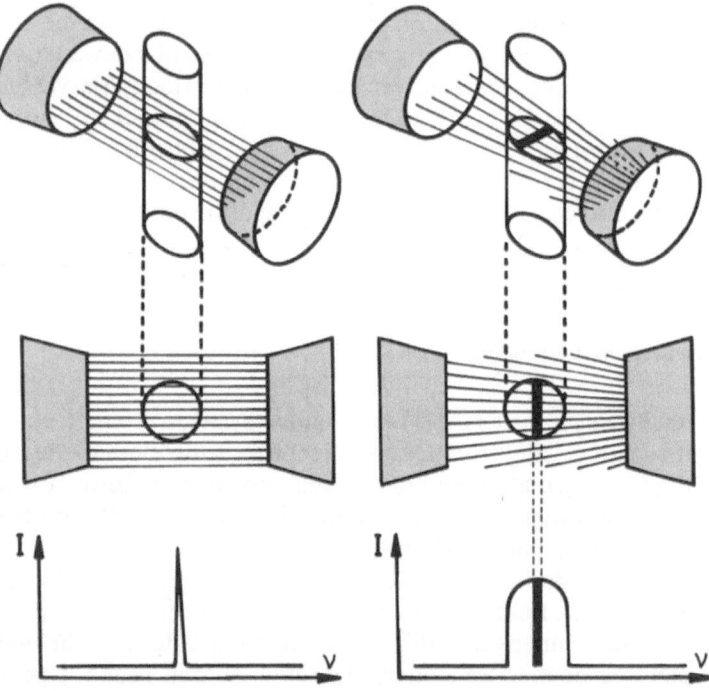

Fig. 56. The physical basis of NMR tomography. In an ordinary NMR experiment of a water-filled tube *(left)* in a homogeneous magnetic field B_o, all nuclei have the same resonance frequency, v_o regardless of their spatial position. For NMR tomography, a nonhomogeneous magnetic field is employed. It is produced by superimposing a linearly magnetic field (gradient) onto the main homogeneous field. The resonance frequencies of the nuclei in this field will vary in accordance with their spatial position. It is this correlation between position and resonance frequency that forms the basis of NMR imaging techniques

portional to the number of hydrogen atoms that are exposed to a magnetic field strength B_0 which exactly meets the resonance condition. Thus, in the presence of the linear magnetic field gradient, signal intensity becomes a direct measure of the water content along a line through the object. This is equivalent to saying that the linear field gradient causes the water concentration to be projected onto the frequency axis of the NMR spectrum. *This correlation between the resonance frequency and spatial position of hydrogen nuclei forms the basis of NMR imaging.*

A cross-sectional image cannot be produced from a single projection; it requires a reconstruction of a serie of projections from a variety of angles.

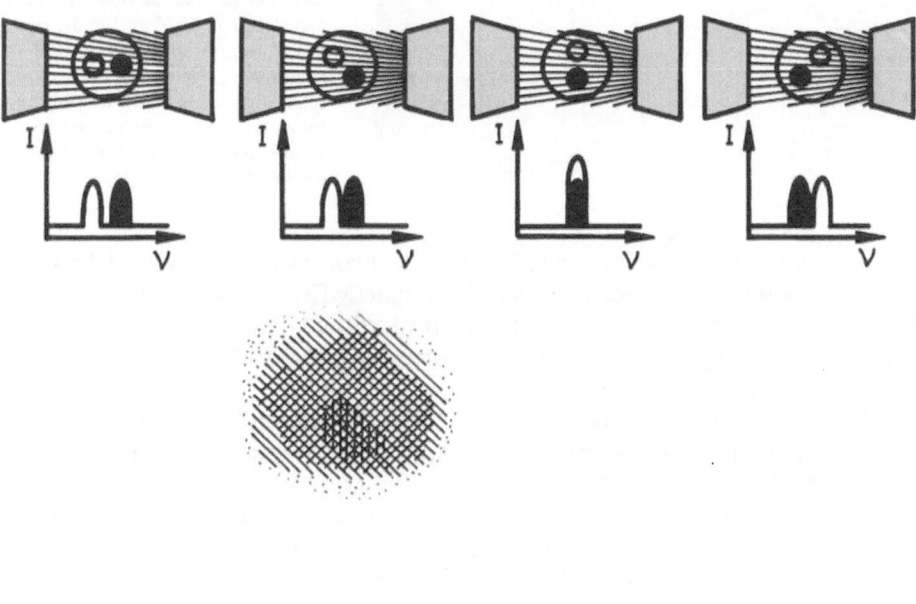

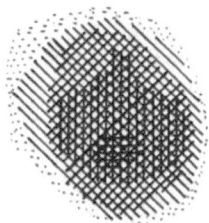

Fig. 57. The principle of NMR tomography. The correlation between spatial position and resonance frequency in a gradient field corresponds graphically to a projection of the water distribution in the object onto the frequency axis of the NMR spectrum. The object or the magnetic field gradient is rotated to obtain multiple projections from which a computer can reconstruct a cross-sectional image of the object. This image of two water-filled glass capillaries (diameter about 1 mm) recorded by Lauterbur in 1973 is the first NMR tomogram. [After Lauterbur PC (1973) Nature 242: 190]

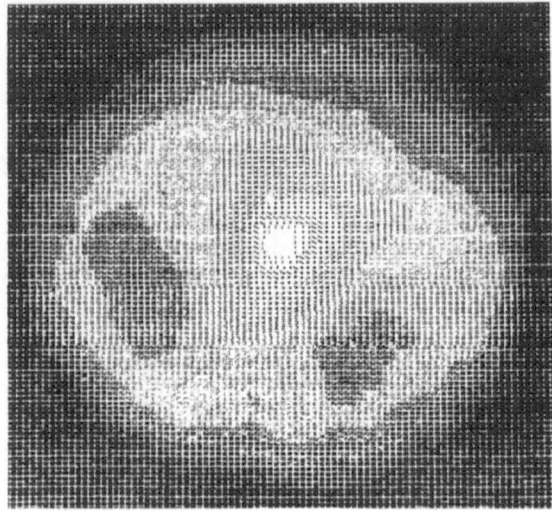

Fig. 58. Axial NMR tomogram of a live mouse. This cross-sectional image through the thorax of an anesthetized mouse was recorded by Lauterbur in 1974 and is the first NMR tomogram of an intact organism. Despite the low resolution, both air-filled pulmonary lobes are clearly visible. [From Lauterbur PC (1974) Pure Appl Chem 40: 149]

Let us assume that the object to be examined consists of two water-filled tubes. As the orientation of the object within the magnetic field is varied, changes appear in the NMR spectrum. From the series of projections obtained by rotating the object within the field, a computer can reconstruct an image of the spatial water distribution within the plane of interest.

This method was used to create the NMR tomogram shown in Fig. 57. This cross-sectional image of two water-filled capillary tubes made by Lauterbur in 1973 was the first NMR tomogram.

Only a year after his first, classic paper was published, Lauterbur recorded the NMR image of a live, anesthetized mouse. Despite the low resolution, the two dark-colored, air-filled pulmonary lobes are clearly visible in the NMR tomogram of the thorax (Fig. 58).

4.2 Image Reconstruction

This section deals with theoretical principles that must be known in order to understand the operation of NMR imagers and to interpret NMR tomograms skillfully. In discussing these principles, it will be necessary to refine the simplified presentations of Chap. 2. Readers who are interested primarily in the practical capabilities of NMR as an imaging technique may skip this chapter for the time being.

In (X-ray) computed tomography, the image plane is defined by the direction and geometry of the X-ray beam. In NMR tomography, on the other hand, an ordinary radiofrequency coil receives a signal, not from a particular plane, but from the entire volume enclosed by the coil. As a result, the signal intensity in the NMR spectrum represents the average concentration of hydrogen nuclei within the enclosed colume, rather than in a specified plane.

A number of highly complex techniques have been devised to solve the problem of "plane selection" in NMR. Below, only the most important methods of producing a sectional image are examined.

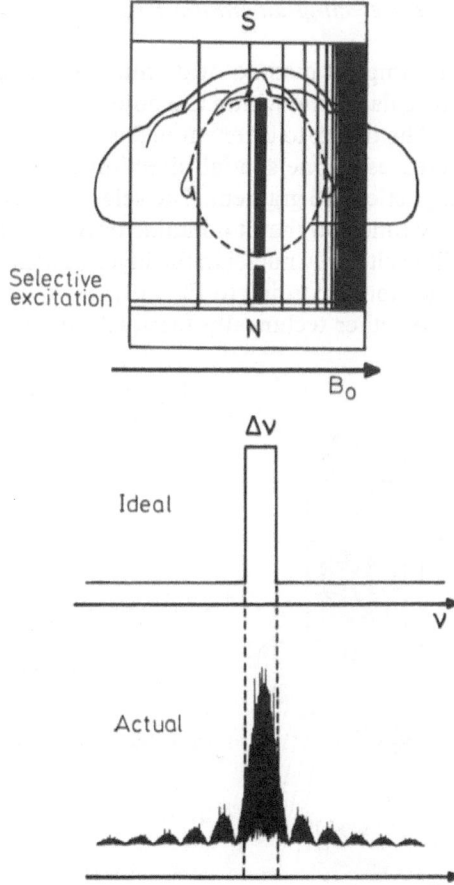

Fig. 59. Defining an image plane by selective excitation. After a linear magnetic field gradient is applied, a particular plane (here sagittal) can be defined by selective excitation, since only the nuclei lying within that plane will be on resonance. The selective radiofrequency pulse is electronically tuned so that its actual frequency distribution (Δv) is narrow and approximates an ideal rectangular shape

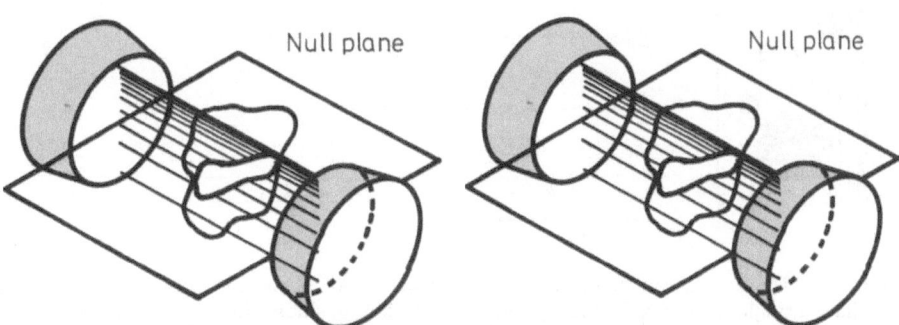

Fig. 60. Defining an image plane by modulating the magnetic field gradient. In biologic samples with a nonhomogeneous structure, it is first necessary to define an image plane. By applying a time-varying auxiliary magnetic field whose polarity is continually reversed (modulation), only nuclei in the "null plane" are exposed to an unchanging field, and only those nuclei will contribute to the measured signal

4.2.1 Selecting an Image Plane

The simplest method of defining an image plane is by the selective excitation of a particular plane within the object. This is illustrated schematically in Fig. 59.

The object to be examined is placed within a magnetic field which progressively increases in one spatial direction owing to the presence of a superimposed linear magnetic field gradient. The selective pulse that is applied to the object is electronically tuned so that it contains only a narrow band of radio frequencies. This pulse will excite only nuclei in the layer in which the magnetic field satisfies the resonance condition according to Eq. (10).

Another technically feasible method of plane selection is based on the use of

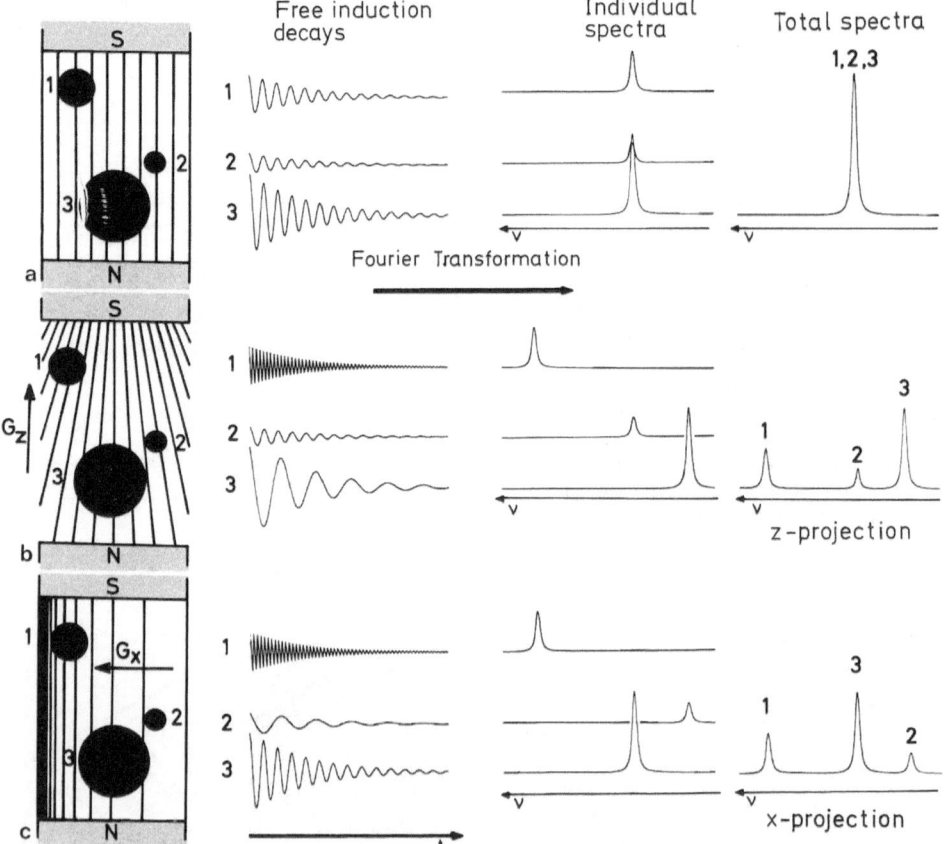

Fig. 61a–c. The acquisition of individual projections. **a** All nuclei in a homogeneous magnetic field have a single resonance frequency, regardless of their position within the object. Hence, the individual free induction decays and spectra from the water-filled tubes of the phantom differ only in their intensity, so the total spectrum contains only one signal. **b** When a z gradient is imposed, the magnetic field strength, and thus the resonance frequency, increases in the order 3-2-1, and the spectrum of the entire object corresponds to a projection of the object onto the z axis. **c** When a x gradient is imposed, the magnetic field strength increases in the order 2-3-1, and the resulting spectrum corresponds to a projection of the object onto the x axis

time-varying magnetic field gradients (Fig. 60). By "modulating" the auxiliary field, i. e., by periodically reversing its direction and magnitude at a particular frequency, only hydrogen nuclei located on the "null plane" at the fulcrum of the oscillations will experience an unchanging magnetic field B_0, and only these nuclei will contribute to signal intensity. Signals from outside this plane will be filtered out as a result of field fluctuations.

4.2.2 Image Reconstruction by Back Projection

After the image plane has been defined, a series of projections are acquired. Rotation of the object in a constant magnetic field gradient, as shown in Fig. 57, is impractical for medical applications. Instead, the object is kept stationary, and the direction of the magnetic field gradient is rotated incrementally around the object within the plane that is to be imaged. At each step around the object, a single projection is taken. This incremental change in the gradient direction is analogous to rotating the X-ray tube in a CT scan (Fig. 61).

Mathematical techniques are available for reconstructing a cross-sectional image of the object from the set of projections obtained. In the simplest case, corresponding signal intensities in different projections can be added together to get a measure of the water content of each picture element (Fig. 62).

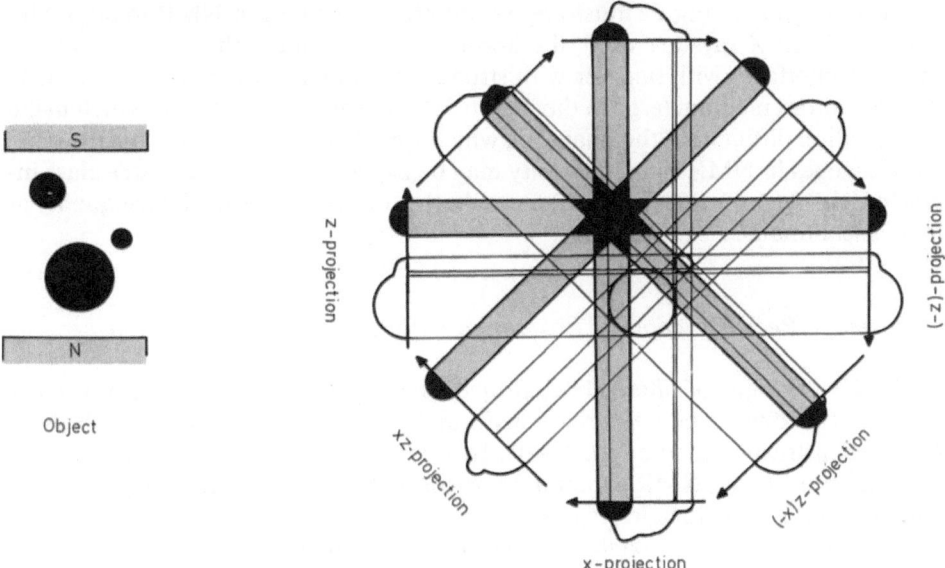

Fig. 62. Image reconstruction by back-projection. The individual projections can be back-projected into an image matrix with the aid of a computer. In the simplest technique, the intensities of corresponding signals are summed together for each picture element, leading to the presence of star-shaped artefacts. In the example shown, the x and z projections (see Fig. 59) are supplemented by projections from other angles. These projections are taken in suitably oriented gradient fields, noting that the gradient direction in the x–z plane can be modified as desired by mixing different x and z gradients together. (For clarity of presentation, the back projection is shown for only one signal from the phantom.)

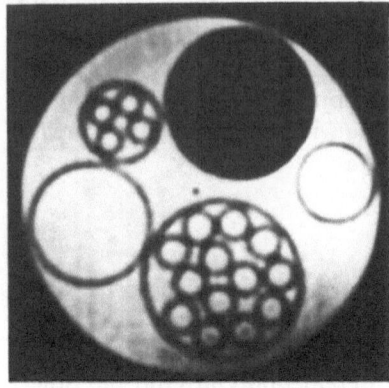

Fig. 63. NMR tomogram of a phantom comprised of empty and water-filled glass tubes. The absence of artifacts at the water-glass interfaces shows the excellent resolving ability of the image reconstruction technique (filtered back-projection) and demonstrates the mastery of the technique over extreme concentration changes. In the gray-scale image, pure water registers as white, while glass and air appear black. The total diameter of the phantom was 23 mm, and the tubes had a wall thickness of 0.2–0.3 mm. (Photo from Bruker, Karlsruhe)

The disadvantage of this technique is obvious: A point in the object appears star-shaped in the back projection. However, the quality of the reconstructed image can be greatly enhanced by suitable mathematical manipulations ("filtered back-projection reconstruction"). Figure 63 illustrates this with the NMR tomogram of a phantom with a nonhomogeneous internal structure. The NMR tomogram, which is free of star-shaped artifacts, documents the excellent image quality that is achieved with this reconstruction method.

The tomogram in Fig. 62 illustrates yet another advantage of NMR tomography over CT. When X-rays are used, the abrupt density changes that occur in places where air interfaces with bone or with strongly absorbent foreign material such as barium contrast medium (e. g., in the digestive tract) can create artifacts which make interpretation difficult. In the phantom, which was also used in Fig. 61, the most extreme changes in NMR signal intensity may be expected to occur at water-glass interfaces (change from 100% to 0% water). As the tomogram shows, image quality at these discontinuities is flawless.

4.2.3 Image Reconstruction by Two-dimensional Fourier Transformation

Unlike the technique of filtered back projection that has proved so useful for X-ray CT scans, two-dimensional Fourier transformation was developed specifically as an image reconstruction method for NMR. To aid in understanding this technique, we shall first consider an NMR experiment performed on a water-filled tube in a nonhomogeneous magnetic field (Fig. 64).

Let us suppose that an excitation pulse is applied to the sample, followed by the application of a magnetic field gradient in the x direction for a time T_E. At the end of T_E, the gradient is switched into the z direction, and the emitted signal, called the "free induction decay" (FID), is recorded.[1] The resonance frequency of the FID de-

---

1 Between the end of the excitation pulse and the start of data collection, an additional inverting pulse is applied to produce a spin echo, and the gradients are switched in a complex fashion. These details, while essential to the practice of NMR tomography, are not necessary for a basic understanding of the two-dimensional NMR technique

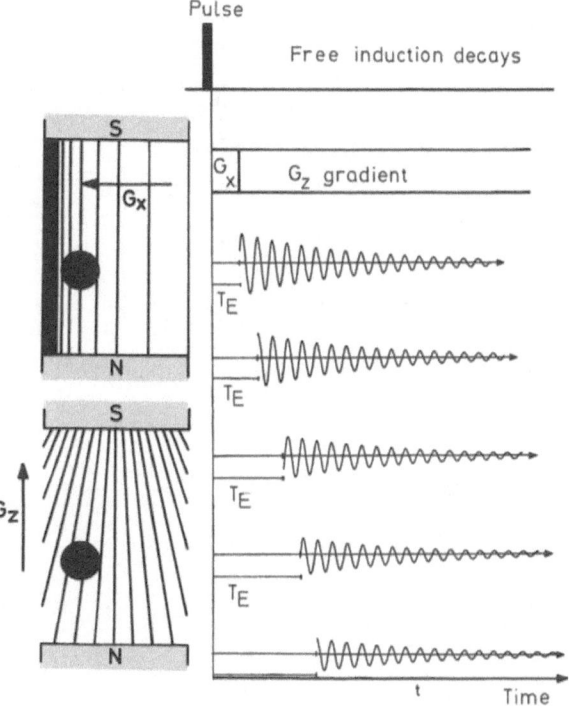

Fig. 64. Recording free induction decays in a magnetic field with switched gradients. Following the excitation pulse, an initial magnetic field gradient is applied in the x direction (G_x) during the waiting period T_E. Then the gradient is switched to the z direction (G_z), and collection of the free induction decay is commenced

pends entirely on the spatial position of the object in the magnetic field with the G_z gradient. At the same time, the signal intensity at the start of the FID (the "phase") is determined by the resonance frequency of the object during the period T_E. Thus, the measured FID contains information on the physical conditions that prevail during the period t (G_z gradient) as well as during the period T_E (G_x gradient).

For two-dimensional image reconstruction, the waiting period T_E during which the G_x gradient is applied is increased in stages. Each FID thus acquired is a function of the two times, T_E and t.

By combining all the free induction decay signals, it is possible to extend the normal one-dimensional Fourier transformation $[I = f(t) \rightarrow I = f(v)]$ to a two-dimensional Fourier transformation $[I = f(T_E, t) \rightarrow I = f(v_E, v)]$, with n FIDs yielding n different spectra (Fig. 65).

The two-dimensional spectrum reflects the physical conditions that prevail during times T_E and t in relation to the v_E and v axes respectively. Because the object was exposed to gradient fields in the x and z directions during the respective times T_E and t, the two-dimensional spectrum corresponds on the v_E axis to a projection in the x direction, and on the v axis to a projection in the z direction. In other words, when the object is exposed to two mutually perpendicular gradients (in this case, G_x

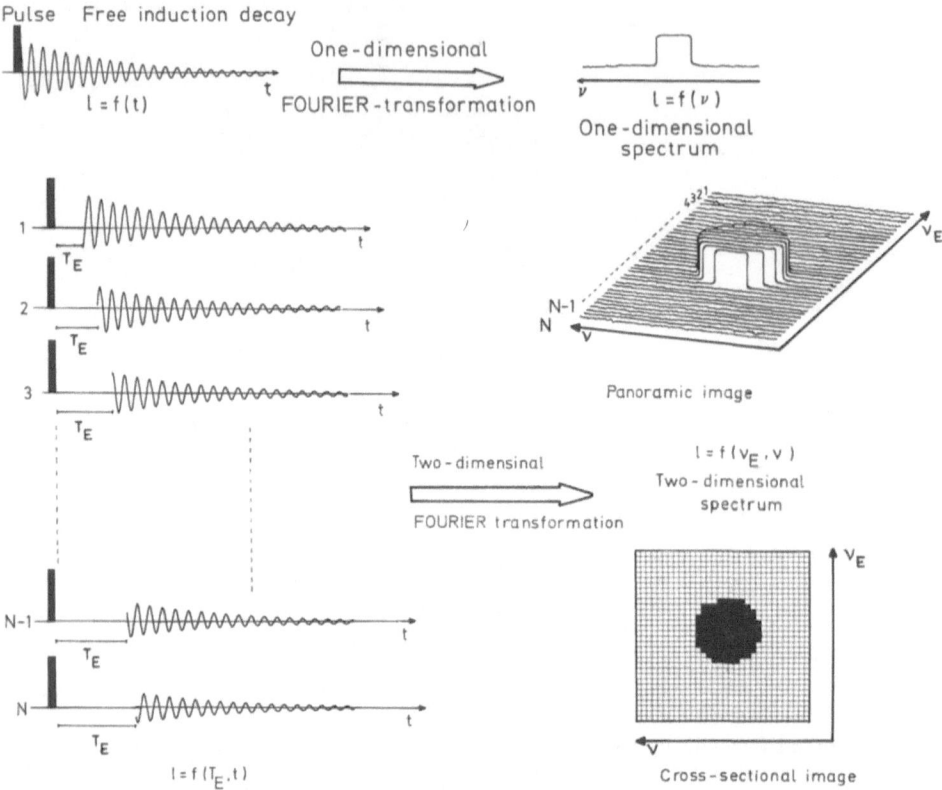

Fig. 65. The two-dimensional Fourier transformation. In an extension of the one-dimensional Fourier transformation, which converts a free induction decay $I = f(t)$ into a spectrum with $I = f(\nu)$, the two-dimensional FT of free induction decays that are a function of two times (t_E and t) yields a spectrum that depends on two frequencies (ν_E and ν). The "two-dimensional" NMR spectrum that is generated may be represented either as a panoramic image which displays the individual spectra in an offset fashion, or as a cross-sectional image in which the intensity of the picture elements is represented by colors or shades of gray

and G_z) during the times T_E and t, the two-dimensional spectrum describes the water distribution in the x–z plane of the object, and the corresponding cross-sectional plot is the NMR tomogram (Fig. 66).

4.3 Measurement of Relaxation Times

It was stated in Section 2.3 that the behavior of biologic tissues on NMR spectroscopy is determined not only by their water content but also by their relaxation times, T_1 and T_2. These relaxation times describe the interaction of tissue water with other cell constituents and so are characteristic of particular tissue types. Because T_1 and T_2 are altered by a number of pathologic states, it is often useful to acquire images which reflect the values of the two relaxation times.

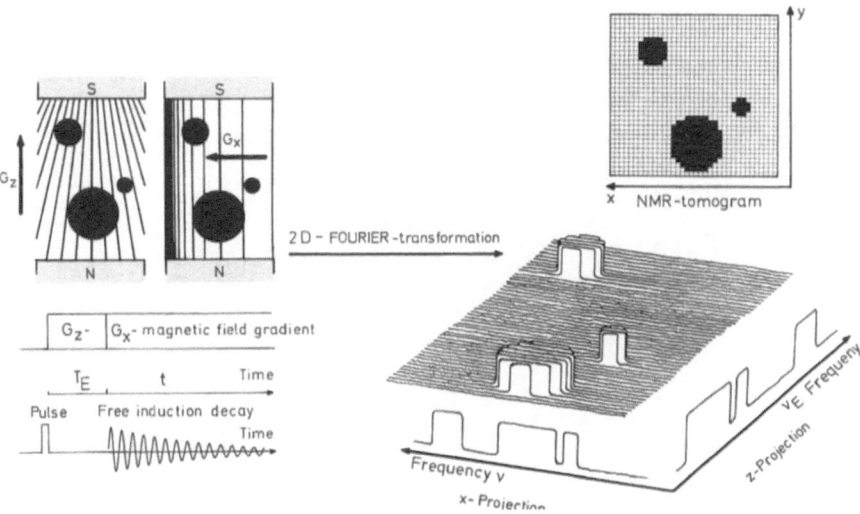

Fig. 66. Image reconstruction by two-dimensional Fourier transformation. The two frequency axes v and v_E in the two-dimensional spectrum reflect the physical conditions that prevail during the times t and T_E. If gradients oriented in different directions (e. g., G_z and G_x) are applied during both time intervals, the spectrum will be an image of the water distribution in the x–z plane of the object

To understand more clearly the techniques needed to measure T_1 and T_2, we shall move from the atomic to the macroscopic scale so that we may examine the behavior not of a single nucleus, but of a large assembly of like nuclear species. This will enable us to describe physical events in classical terms that are more amenable to schematic illustration.

When a great many nuclear magnets are assembled their magnetic properties are added together. In the absence of an external magnetic field (see Fig. 5), the individual magnets point in random directions, with the result that their net external magnetic effect averages to zero. Water, for example, is completely nonmagnetic externally. When a magnetic field is applied, however, the nuclear magnets become aligned either parallel to the field or antiparallel to it. Because the parallel alignment has a somewhat lower energy, it is occupied more frequently than the antiparallel alignment. When summed over all the nuclei, this small excess of parallel nuclei leads to a net magnetization whose *magnitude* is determined by the *population difference* (Fig. 67).

In the NMR experiment, a pulse of radiofrequency energy is applied to the sample by means of a transmitter coil mounted at right angles to the magnetic field direction. For an instant this exposes the nuclear magnets to a second field, B_1. According to the rules of classical physics, this leads to a rotation (precession) of the net magnetization M_0 in the y–z plane (Fig. 68).[1]

The angle through which the net magnetization is tipped, continues to increase as long as the radiofrequency energy is applied. A "90° pulse" is one whose pulse

---

1 For didactic reasons I take the liberty of mixing the rotating and the stationary coordinate systems. A physically correct but mathematically involved interpretation is given in Appendix A

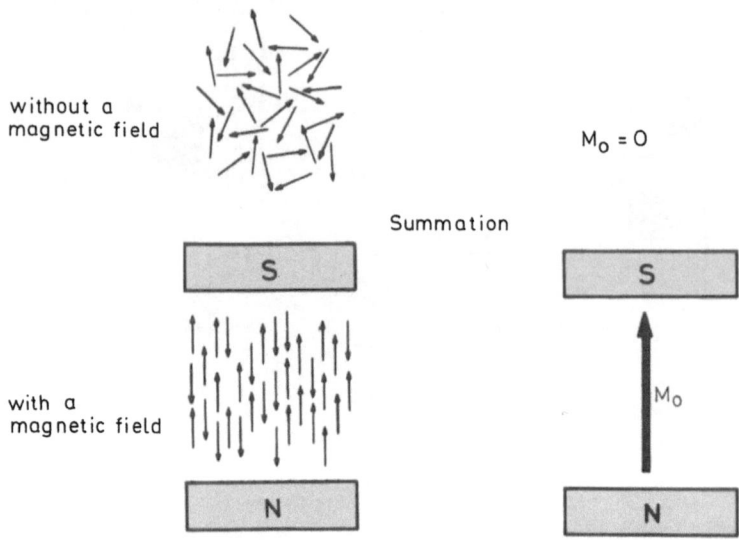

without a
magnetic field

$M_0 = 0$

Summation

with a
magnetic field

M_0

Fig. 67. The magnetic properties of atomic nuclei. On the atomic scale, the quantum theory describes the behavior of individual nuclei in a magnetic field. The properties of a very large number of particles, however, can be described in terms of classical physics. From this viewpoint the net magnetization M_0 represents the sum of all the individual magnetizations of the nuclei. Because the lower energy level (parallel alignment) is more heavily populated, there is a net magnetization directed parallel to the external field

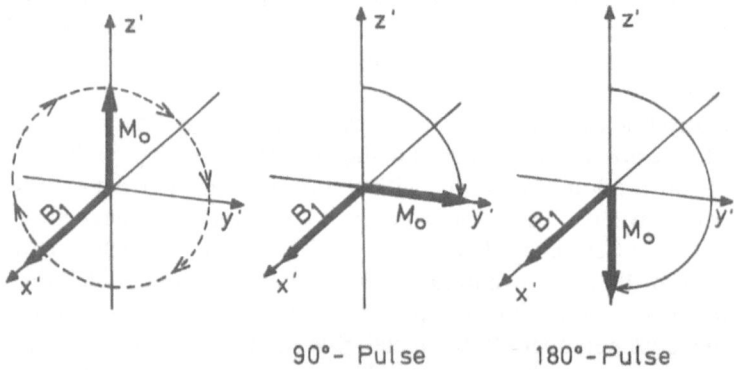

90°- Pulse   180°- Pulse

Fig. 68. Effect of the excitation pulse. When an excitation pulse is applied, the nuclear magnets are subjected to an additional field, called B_1. According to the laws of classical physics, this causes the magnetization M_0 to rotate, or "precess," about the B_1 axis. At the end of the excitation pulse, the net magnetization may be tipped by 90° or 180° (thus pointing in the y' or (–)z' direction), depending on the length of the pulse

length tips the net magnetization in the $+y'$ direction, causing it to precess at right angles to the z axis. By doubling the pulse length (180° pulse), the net magnetization can be turned completely over so that it is directed opposite the main field. Quadrupling the pulse length (360° pulse) returns the net magnetization to its original position.

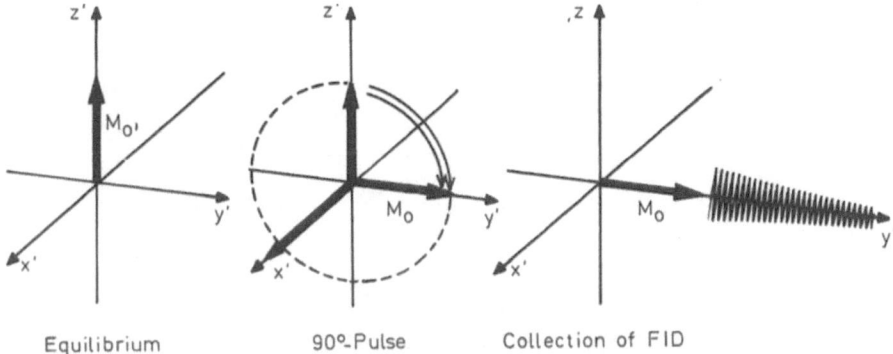

Fig. 69. The pulsed NMR experiment. A 90° excitation pulse is applied, tipping the net magnetization in the y' direction. The receiver coil, which is sensitive in this direction, picks up the free induction decay, which is Fourier-transformed to yield the NMR signal

In this expanded model, the events occurring in a normal NMR experiment may be described as follows (Fig. 69):

First a 90° radiofrequency pulse of several microseconds' duration is applied to the sample. This rotates the net magnetization from the z direction to the +y direction. After the excitation pulse ends, the free induction decay of the net magnetization in the y direction is recorded. The FID is then Fourier-transformed to obtain the NMR spectrum.

After the 90° pulse ends, the net magnetization eventually returns to its original, equilibrium position, i.e., it returns from the +y direction to the z direction. This process, called relaxation, has two mutually independent components: a change of magnetization in the y direction, and a change of magnetization in the z direction (Fig. 70).

As relaxation proceeds, the +y' component of M_0 decays to zero (transverse relaxation), while the z' component increases from zero to M_0 (longitudinal relaxation). The relaxation times T_1 and T_2, which were introduced phenomenologically in Chapt. 2, are identical to the longitudinal (T_1) and transverse (T_2) relaxation times defined here.

4.3.1 Spin-Lattice Relaxation Time T_1

The signal intensity of the FIDs in a series of 90°-read pulses (the "saturation-recovery" pulse sequence) demonstrates the relationship, defined in Eq. (6), between the repetition time T_R and the relaxation time T_1. With a long repetition time ($T_R \gg T_1$), signal intensity depends entirely on the water content of the volume being examined. As T_R is shortened, signal intensity is increasingly influenced by the relaxation time T_1. In principle, then, the information content of the NMR image can be altered by manipulating T_R. The spin-lattice relaxation time can be determined from several tomograms with different T_R values, but a better way is to use the "inversion recovery" (IR) technique. In this technique one first applies a 180° inverting pulse that tips the net magnetization from the (+)z' to the (−)z' direction (inver-

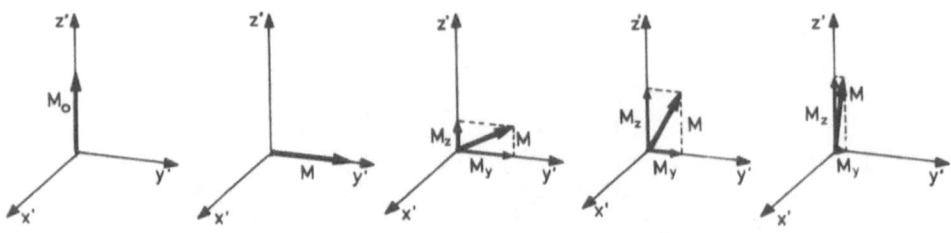

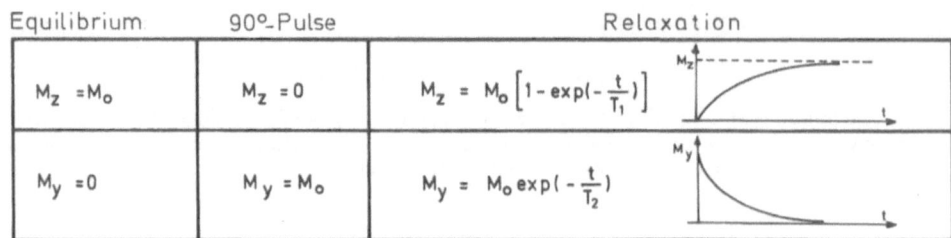

Fig. 70. The relaxation process following a 90° pulse. The 90° pulse tips the net magnetization from the z' to the +y' direction. The subsequent recovery process is described by two relaxation times: T_2, which describes the decay of the magnetization component in the y' direction (M_y) from M_o to the equilibrium value zero; and T_1, which characterizes the return of the magnetization component in the z direction (M_z) from zero to the equilibrium value M_o

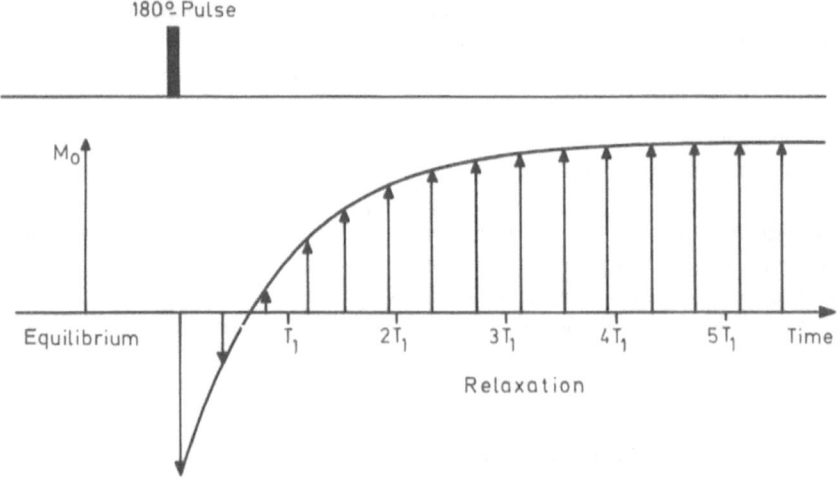

Fig. 71. The relaxation process following a 180° pulse. The 180° pulse tips the net magnetization from the +z to the −z direction. During subsequent relaxation the magnetization tends to resume its original value and direction in a process described by the relaxation time T_1

sion). The return of the magnetization to equilibrium along the z axis is shown schematically in Fig. 71. However, there is no way to observe relaxation along the main magnetic field, because the receiver coil can measure components in the y' direction only. Thus, in order to detect magnetization along the z axis with the receiver coil it is necessary to apply a 90°-read pulse after an interval T_I, which will tip the magnetization in the y' direction (Fig. 72 a).

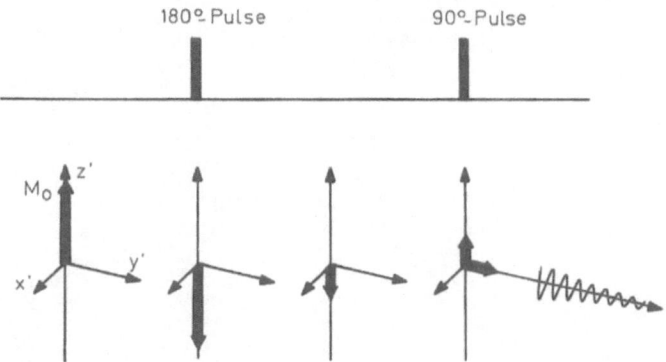

Fig. 72. The inversion-recovery (IR) experiment. Following a 180° inverting pulse, the magnetization returns from the $-z$ to the $+z$ direction. Because the receiver coil can detect magnetization in the y direction only, the T_1 relaxation process along the z axis can be observed only by applying a 90° "read" pulse after an interval T_I which tips the magnetization in the y direction

The pulse pattern $180° - T_I - 90°$ is called the inversion recovery sequence. From two or more such measurements, the T_1 values in a tissue section can be determined. By combining IR with a normal NMR tomogram, a T_1 NMR tomogram can be generated in which different T_1 values are displayed in either a color or gray-scale format (Fig. 73).

For clinical diagnostic purposes, the measurement of T_1 values themselves is usually less important than tissue discrimination. Fig. 74 shows schematically how this discrimination is accomplished for the inversion recovery measurement of two signals having equal intensities but different T_1 values.

The *intensity difference,* which ultimately determines image contrast, depends to a critical degree on the delay time T_I between the 180° pulse and the 90° pulse.

Figure 75 shows the results of IR measurements in axial NMR tomograms of the human head. Owing to the characteristic time dependence of the recovery process (Fig. 71), the IR signal changes its sign after an interval equal to 0.69 T_1, causing a reversal of contrast in the IR image.

4.3.2 Spin-Spin Relaxation Time T_2

In principle, it should be possible to determine the spin-spin relaxation time T_2, representing the change of magnetization in the y' direction, directly from the FID of the NMR experiment, since the emitted signal decays to zero with T_2 as its time constant.

FID in a homogeneous field: $I = I_0 \exp(-t/T_2)$ (9 b)

However, Eq. (9) is valid only if the main magnetic field is perfectly homogeneous. The imperfections that are invariably present in magnetic fields, especially when gradients are imposed, causes relaxation to proceed at a considerably faster rate

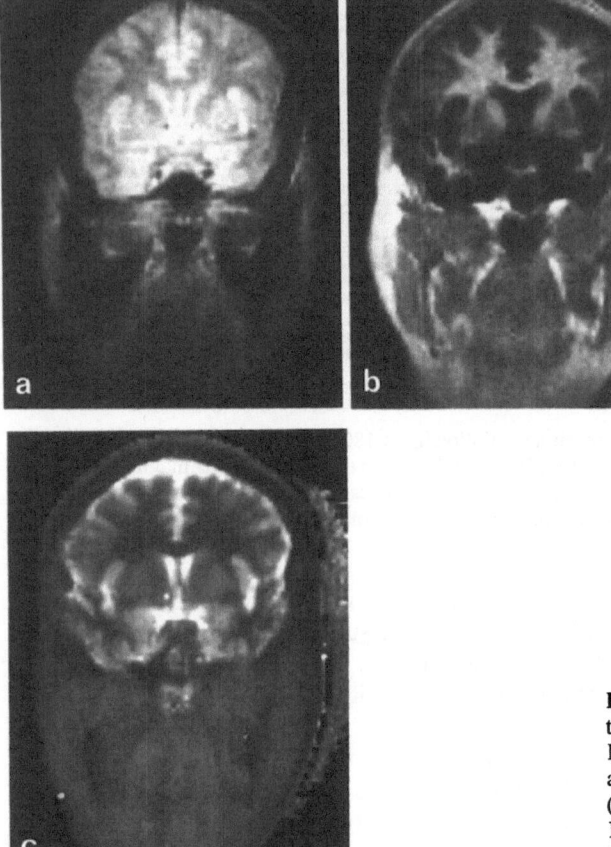

Fig. 73 a–c. Coronary NMR
tomograms of the human head.
From an ordinary NMR image (**a**)
and an inversion-recovery image
($T_1 = 400$ ms) (**b**), a richly detailed T_1
NMR tomogram (**c**) can be generated.
(Photo from Siemens, Erlangen)

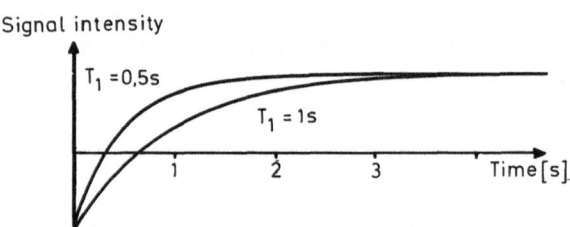

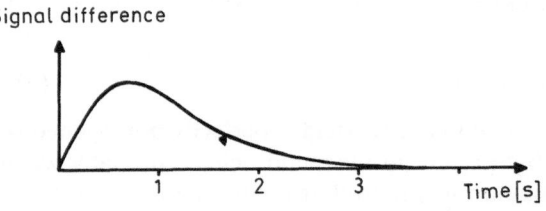

Fig. 74. Recording of two signals
with different T_1 relaxation times by
the inversion-recovery technique.
Following the 180° inverting pulse,
two equally strong signals with
different T_1 relaxation times return
to equilibrium in accordance with
their respective T_1 values, during
which time the intensity difference,
which determines image contrast,
passes through a maximum

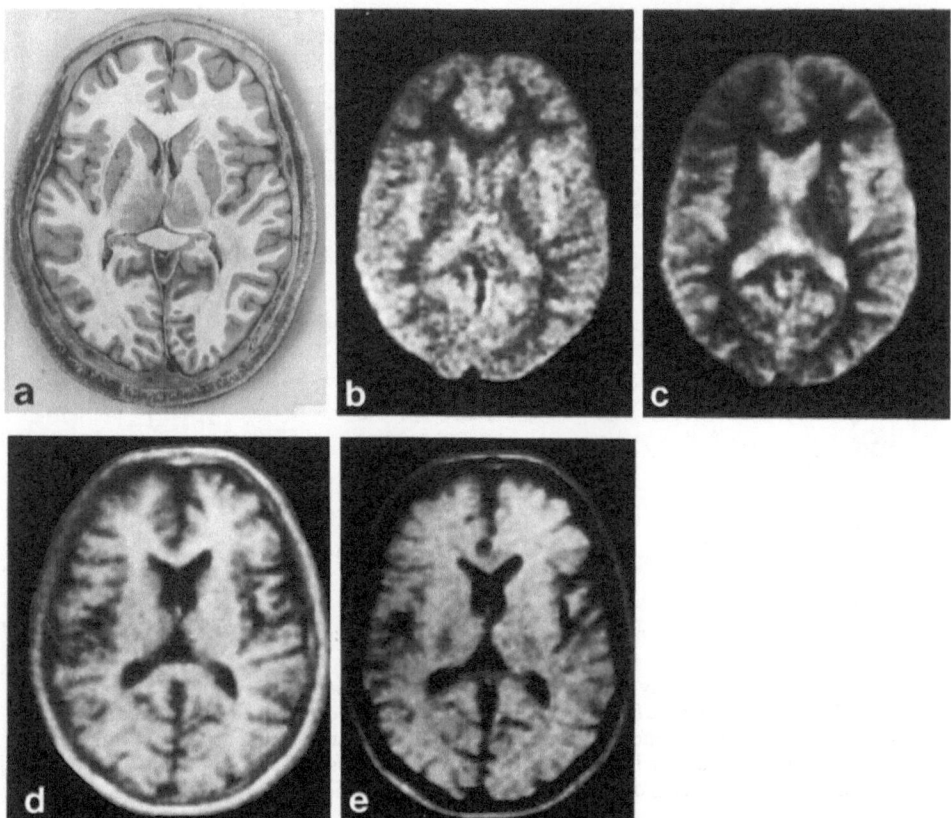

Fig. 75 a–e. Axial IR NMR tomograms of the human head. **a** Anatomic section. **b–e** Inversion recovery images (180°-T_I-90°). **b** $T_I = 100$ ms, **c** $T_I = 200$ ms, **d** $T_I = 400$ ms, **e** $T_I = 800$ ms. With short intervals T_I between the 180° inverting pulse and the 90° read pulse, the ventricles of the brain register as dark (**b, c**). With a longer T_I, the ventricles appear light. As T_I is increased, contrast is heightened between the white and gray matter of the brain and is maximal at about 400 ms. [From Ziedses des Plantes BG Jr et al. (1983) In: Wende S, Thelen M (eds) Kernspin-Tomographie in der Medizin. Springer, Berlin Heidelberg New York Tokyo]

(shorter relaxation time). The cause of this accelerated relaxation lies in the different resonance frequencies of the hydrogen atoms in the nonuniform magnetic field (Fig. 76).

When all the different free induction decays of the nuclei are added together, they yield a summation FID which has a new, shorter relaxation time, designated T_2^* ("T two star")

FID in a nonhomogeneous field: $I = I_0 \exp(-t/T_2^*)$ (10)

Despite the apparent shortening of the relaxation time of the entire sample from T_2 to T_2^*, the individual nuclei continue to relax in accordance with their natural T_2 values. Only T_2^* is directly measured, however. In human brain tissue, for example, a relaxation time of only 5–10 ms is observed in a typical tomographic experiment, despite the fact that the actual T_2 value ranges between 100 and 150 ms. This shows

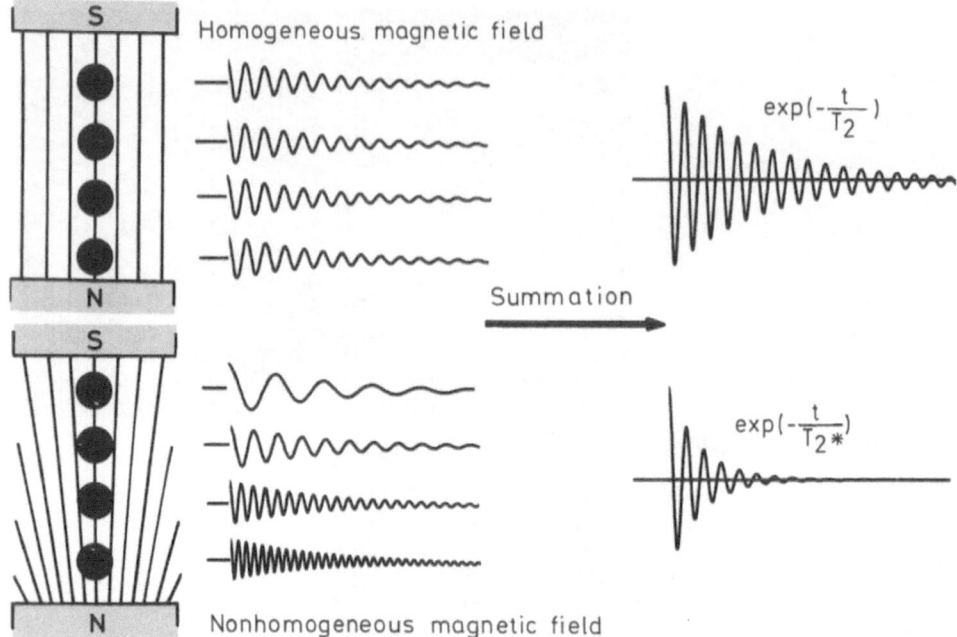

Fig. 76a, b. Free induction decay in homogeneous and nonhomogeneous magnetic fields. **a** In a perfectly homogeneous magnetic field, the rate of signal decay is determined by the natural relaxation time T_2. **b** In a nonhomogeneous magnetic field, the resonance frequency varies with spatial position. Summation over all the individual free induction decays, which vary slightly in their frequency, yields a total FID that is described by the shorter relaxation time T_2^*

that the value of T_2^* is determined almost entirely by nonhomogeneities of the magnetic field, and that T_2^* tells us nothing about the true relaxation time T_2 of the tissue.

By using an ingenious technique called the "spin-echo" experiment, it is nonetheless possible to determine the true spin-spin relaxation time of the tissue by eliminating the effects of magnetic field nonhomogeneities. The irradiation sequence in this technique consists of an initial 90° excitation pulse, followed by a 180° inverting pulse after a waiting period $T_E/2$ (Fig. 77).

In a perfectly homogeneous magnetic field, the only effect of the 180° pulse is to cause an abrupt change of sign in the FID. In a nonhomogeneous field, however, the inverting pulse leads to an increase in the observed signal by summation over all the individual FIDs, leading to the presence of a "spin echo".

Prof. Ray Freeman of Oxford has suggested an excellent analogy to aid in understanding this physical phenomenon. Prof. Freeman likens the spin-echo experiment to a rather curious horse race. After the starting gun is fired (90° pulse), all the horses (spins) start running in accordance with their ability (resonance frequency). Within a short time the strongest horses (high resonance frequency) are far in the lead, while the weakest horses (low resonance frequency) have barely left the starting gate. After a time $T_E/2$, a second shot is fired (180° pulse), whereupon all the horses (spins) reverse their direction and run back for the gate. After a further interval $T_E/2$, *all* the horses, regardless of their speed (resonance frequency), reach the starting gate at the *same time* (spin echo).

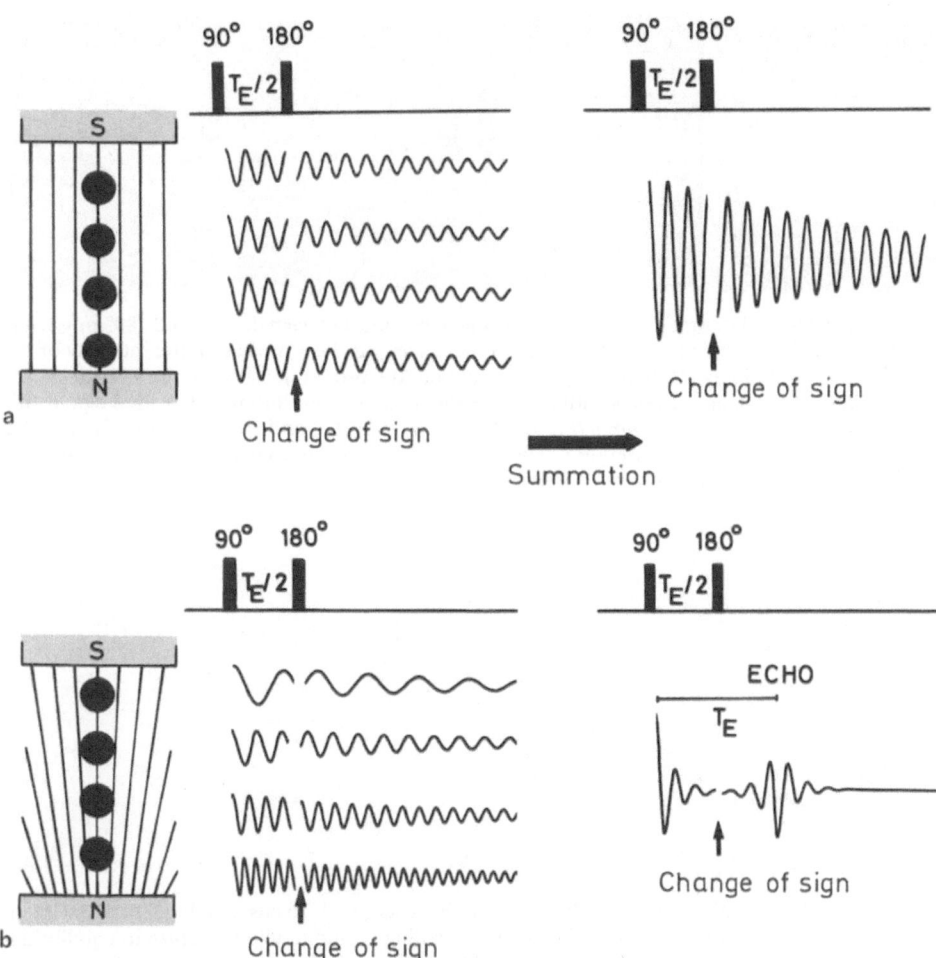

Fig. 77 a, b. The spin-echo experiment. The addition of a 180° inverting pulse to the homogeneous magnetic field (**a**) causes an abrupt change of sign in the emitted signal, while all other properties of the free induction decay remain unchanged. In a nonhomogeneous field, this applies to each and every FID signal (**b**). In contrast to **a**, summation of all the FIDs, which differ slightly in their resonance frequencies (but not in T_2!) due to field imperfections, leads to a gradual increase in measured signal intensity. After a time the intensity of the "spin echo" becomes maximum. The intensity of the spin echo is determined entirely by the natural T_2 relaxation time

The spin echo which ultimately appears contains all the information of the original free induction signal, except that the intensity of the spin echo is reduced by the factor $\exp(-T_E/T_2)$. Thus, the intensity of the spin echo is determined exclusively by the intrinsic T_2 value of the sample. In NMR tomograms that are generated from spin echoes, both the T_2 values and the proton density of the tissue contribute to the final image (Fig. 78).

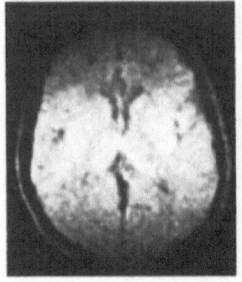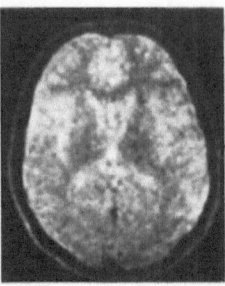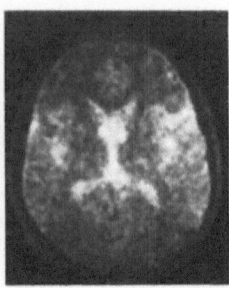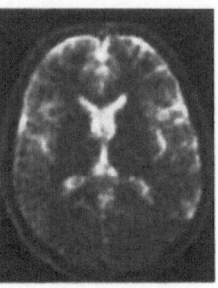

Fig. 78. Axial spin-echo tomograms. By varying the delay time between the 90° and 180° pulse, it is possible to heighten the contrast between different tissues. As T_E is increased, the intensity of the echo signal is increasingly influenced by T_2, so that both the proton density and the value of T_2 are reflected in the final image. Tissues with the same water content but different T_2 values can be differentiated by this technique. *From left to right:* $T_E = 50$, 100, 150, and 200 ms. [From Ziedses des Plantes BG Jr et al. (1983) In: Wende S, Thelen M (eds) Kernspin-Tomographie in der Medizin. Springer, Berlin Heidelberg New York Tokyo]

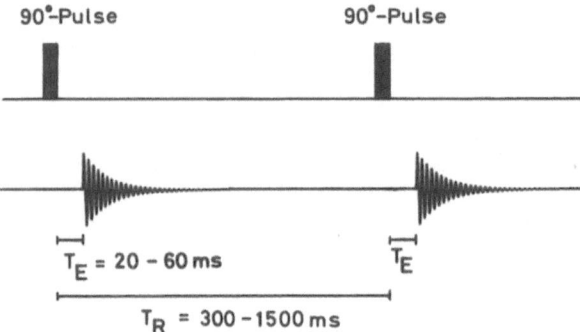

Fig. 79. Time diagram of a simple NMR tomographic experiment. To ensure adequate signal intensity, each NMR experiment must be followed by a waiting period to allow a return to equilibrium (cf. Fig. 19). The numerical data illustrate the unfavorable relationship between signal-acquisition time and waiting time

4.4 Data Acquisition Strategies in NMR Tomography

As stated earlier, the NMR experiment is inherently insensitive due to the small population difference between the parallel, low-energy state and the antiparallel, higher-energy state. To produce an image of satisfactory quality, therefore, signals must be acquired from a given volume many times and then summed in a computer. However, because the return to equilibrium during spin-lattice relaxation following an excitation pulse and FID collection is a slow process, the NMR experiment cannot be repeated right away (cf. Fig. 19). The time diagram of a typical NMR tomographic experiment is shown in Fig. 79. This technique is inefficient.

Following a 20-ms period of actual data acquisition, there is a relatively long (> 1 s) and unproductive waiting interval before the next pulse can be applied. Various manufacturers and research groups have developed methods for improving the time efficiency of NMR imaging by making use of the long waiting period.

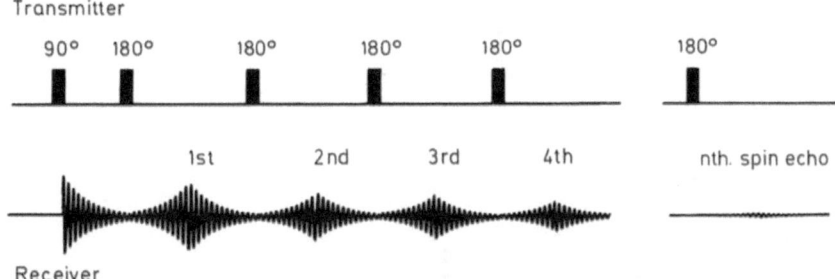

Fig. 80. Multiple spin-echo or Carr-Purcell experiment. After a 90° excitation pulse, 180° inverting pulses are applied to evoke multiple spin echoes, each of which contains all the image information and can be used to produce a tomogram

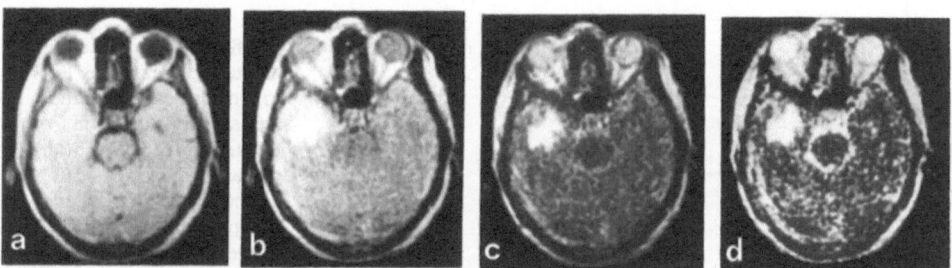

Fig. 81 a–d. Axial spin-echo NMR tomograms of a 42-year-old woman with cerebral metastasis following a mastectomy. History: Following a mastectomy and two local recurrences in the area of the amputation scar, the patient complained of hypoesthesia in the area of the 1st and 2nd divisions of the trigeminal. This was accompanied by complete abducens paralysis on the right side and partial oculomotor paralysis. A CT scan with contrast medium showed no definite abnormalities. This was followed by an NMR tomographic examination using the multiple spin-echo technique, in which a total of 24 spin echoes were recorded, with consecutive series of six echoes being processed into an image. **a** The first tomogram (1st–6th spin echoes) essentially represents a proton-density image. **b–d** The remaining tomograms, produced from the three subsequent series of spin echoes, were made with increasing T_E values and reflect the increasing influence of the T_2 values of the tissue on image contrast. While the only striking feature in the proton-density image (**a**) is a very slight increase in proton density in the left temporal lobe, the subsequent tomograms, which are increasingly T_2-weighted, clearly demonstrate a space-occupying lesion. [After Friedburg H and Bockenheimer S (in press) Radiologe; tomograph from Bruker, Karlsruhe]

In one method, an initial 90° excitation pulse is applied, the FID is collected, and then a sequence of 180° inverting pulses is applied to evoke multiple spin echoes (the "Carr-Purcell" sequence) (Fig. 80).

The second half of each echo is recorded. Each echo contains a full set of image information, with T_2 making an ever greater contribution to image intensity as time t increases. After the last echo ends, it is still necessary to wait until the population difference returns to equilibrium, but because several FIDs can be recorded per cycle, the overall efficiency of the imaging process is greatly improved.

The diagnostic example shown in Fig. 81 documents the fact that the Carr-Purcell technique not only improves time efficiency, but also permits a better tissue discrimination owing to the longer T_E intervals and the consequently greater influence

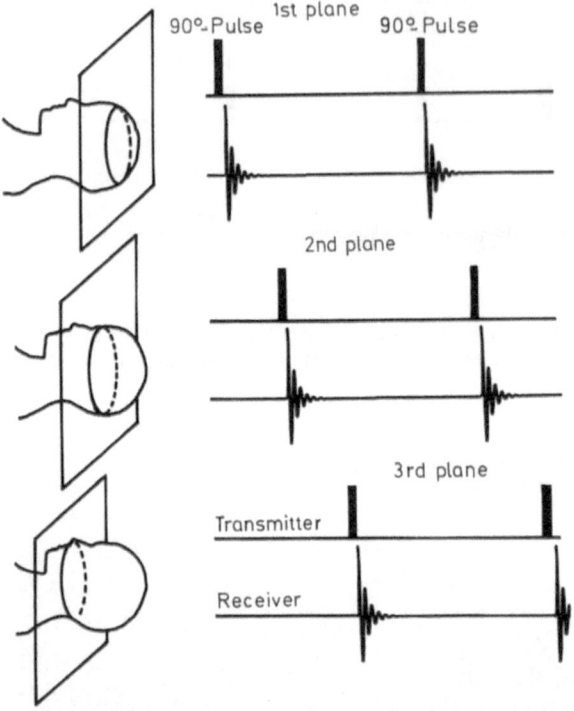

Fig. 82. The multiplanar excitation technique. Following selective excitation and signal acquisition from the 1st image plane, additional image planes can be selectively irradiated and imaged while the previous plane is returning to equilibrium

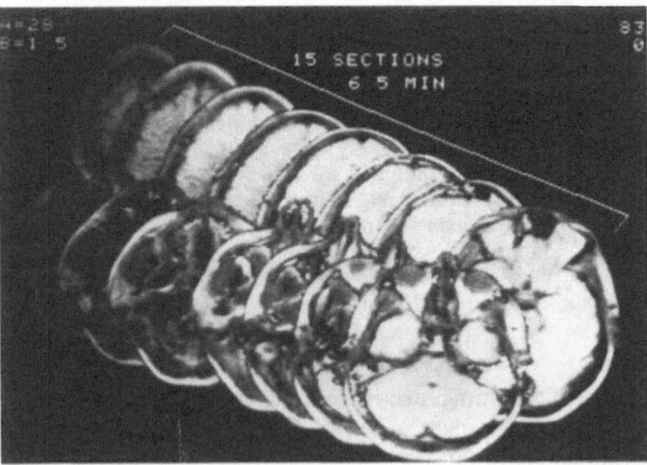

Fig. 83. Application of the multiplanar technique. The axial NMR tomograms from a total of 15 different image planes in the human head can be acquired within a total of 6.5 min. [Photo from Prof. L. Kaufman, Diasonics, San Francisco]

of T_2 values on image contrast. Computer analysis of the tomogram permits the T_2 relaxation times of the tissues under examination to be determined.

While the brain tissue of healthy control subjects has T_2 values between 115 and 130 ms, the T_2 of metastases from breast carcinoma is 280 ms. This doubling of T_2 makes diseased tissue easy to identify in the spin-echo image.

Another way to increase the efficiency of NMR tomography is the simultaneous imaging of multiple sections (Fig. 82). A selective pulse is used to interrogate a given image plane (cf. Fig. 59), so that nuclei located outside this plane will not be stimulated. By applying selective pulses to all the image planes in a sequential fashion the signal from one plane can be recorded while the other planes are returning to equilibrium. The time diagram of the multiplanar imaging technique is shown in Fig. 82. A practical application is illustrated in Fig. 83.

Besides these two methods, a number of other interesting strategies have been devised for making NMR imaging more efficient. A discussion of these techniques would be beyond the scope of the present text (see references at end of chapter).

4.5 The NMR Tomogram

4.5.1 Tissue Properties Which Affect the Image

In the CT scan, the relative brightness levels in the picture reflect the X-ray absorption coefficient of the tissue and can be referred to a *single* standard scale (the Hounsfield scale). By contrast, the relative intensity of NMR images is determined by various tissue properties and by the imaging parameters that are selected. Based on the principles outlined in Sect. 4.2, image intensity in the NMR tomogram is given by the equation

$$I = N(H) \cdot \exp(T_E/T_2) [1 - \exp(T_R/T_1)] \tag{11}$$

Tissue properties	Imaging parameters
$N(H)$, water content;	T_E, delay time between excitation and signal acquisition;
T_1 spin-lattice relaxation time;	
T_2 spin-spin relaxation time.	T_R, delay time between data acquisitions.

Equation (11) forms the basis of NMR tomography and states mathematically the effect of the characteristic tissue properties $N(H)$, T_1, and T_2 and of imaging parameters on the final image intensity. The relationship between NMR properties and NMR image intensity for a given set of imaging parameters is shown in Table 4.

Tissue Water Content
Tissue water content has a major effect on the intensity of picture elements in an NMR tomogram. Water content not only varies among different tissues, but may also vary in the same tissue as a result of nutrition, climate, or drug therapy. Moreover, the water content of human tissues tends to decrease with age, by an average of 2.5%/decade. The water contents of different human tissues are listed in Table 5.

Table 4. Correlation between image brightness and tissue properties

Tissue property	Image brightness
Water content	Tissue containing large amounts of water appears light
Spin-lattice relaxation time T_1	Tissue with long T_1 appears dark
Spin-spin relaxation time T_2	Tissue with long T_2 appears light

Table 5. Water content of various tissues in man

Tissue	Water content (%)
Skeletal muscle	79
Heart	79–80
Liver	71
Kidney	81
Spleen	79
Brain	
– white matter	84
– gray matter	70–74
Epidermis	65
Teeth	3–10
Placenta (20–40 weeks)	87

The diagnostic use of NMR tomography relies on the fact that numerous pathologic processes are associated with a change of water content in the involved tissues. The water contents of various normal and malignant tissues in the rat are shown in Fig. 84.

Relaxation Times T_1 and T_2

The relaxation times are an indirect measure of the interactions that occur between water molecules and other cell constituents. Due to the complex way in which water may be bound to the various cell components it is not possible to describe the relaxation mechanisms in tissues in quantitative, theoretical terms. For simplicity, we may classify the water molecules in biologic tissues into two types: "free" and "bound". The "free" water is by far the more abundant and behaves like ordinary water in terms of its NMR properties. The "bound" fraction, which ranges between 5% and 10%, consists of water molecules that are bound rather firmly to the surface of macromolecules (cell membranes, proteins, etc.) and are therefore far less mobile than their free counterparts. This is illustrated schematically in Fig. 85.

Owing to their relative immobility, the bound water molecules can transfer absorbed radiofrequency energy to the environment much more rapidly than can the free water molecules. As a result, the T_1 relaxation time of bound water is up to 1000 times shorter than that of free water.

Free and bound water do not produce separate signals in the NMR spectrum, only a single, averaged signal whose T_1 relaxation time is determined by the average values of the free and bound water:

$$1/T_1 = [1/T_1(b)] \cdot P_b + [1/T_1(f)] \cdot P_f \qquad (12)$$

where $T_1(b)$ and $T_1(f)$ are the relaxation times and P_b and P_f are the percentage contents of bound and free cellular water respectively in the tissue examined.

Because $T_1(f)$ is much greater than $T_1(b)$, Eq. (12) may be simplified to an approximation:

$$T_1 \approx T_1(b) \cdot (1/1 - P_f) \qquad (13)$$

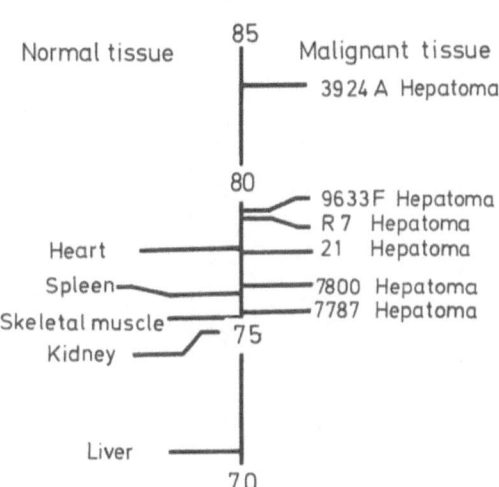

Fig. 84. Water content of various normal and malignant rat tissues. [After Hollis DP and Eggleston LA (1975) J Natl Cancer Inst 54: 1469]

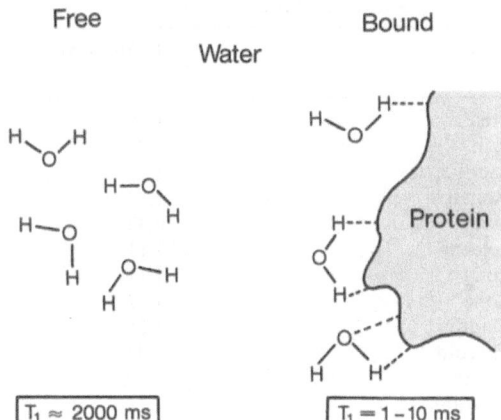

Fig. 85. The relaxation of "free" and "bound" cellular water. Due to the limited mobility of water molecules that are bound to the various cell components, bound water has a shorter T_1 relaxation time than water in the free state does

Thus, the measured T_1 value of a tissue is determined primarily by a tissue-specific constant $T_1(b)$ which depends on the interaction of the "bound" water with cell components. T_1 is further determined by the relative content of "free" water in the volume of interest. Because a rise of total water content mainly increases the relative abundance of free water, *the T_1 value of a tissue increases with its water content*. This fact has been confirmed experimentally for both T_1 and T_2 in various normal and malignant rat tissues.

As the slopes of the lines in Fig. 86 indicate, a 1% increase in tissue water content leads to an approximately 8% increase in relaxation time. This sensitive response of relaxation time to variations in tissue water content underlies the diagnostic value of NMR tomograms that are based on tissue relaxation properties.

The increase in the T_1 relaxation time of tissues which accompanies many disease states is a major factor in the clinical importance of NMR imaging (Table 6).

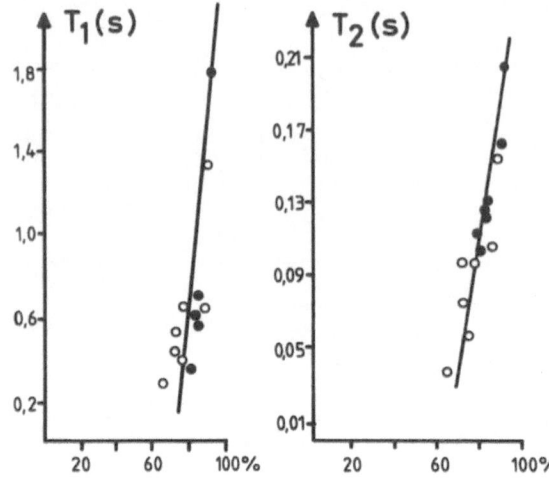

Fig. 86. Correlation between the relaxation times T_1 and T_2 and the water content of tissue. The comparison between miscellaneous (O) and malignant (●) rat tissues shows that the relaxation times T_1 *(left)* and T_2 *(right)* increase with the tissue water content. [After Kiricuta IO and Simplaceanu V (1975) Cancer Res 35: 1164]

Table 6. Relaxation times of normal and tumorous tissues at 0.051 tesla [After Damadian R (1980) Phil Trans R Soc Lond B 289: 489]

Tissue	T_1 (tumorous) (s)	T_1 (normal) (s)
Breast	1.08	0.36
Skin	1.04	0.61
Esophagus	1.04	0.80
Stomach	1.23	0.76
Liver	0.83	0.57
Spleen	1.11	0.70
Lung	1.11	0.78
Bone	1.02	0.55
Bladder	1.24	0.89
Thyroid	1.07	0.88
Uterus	1.39	0.92
Prostate	1.11	0.80

Although the cause of the increased relaxation time in many malignant tissues is not yet fully understood, it must be due in large part to an abnormally high water content. Relaxation time is also prolonged in other types of fast-growing tissue (e. g., normal fetal tissue) as well as in many benign pathologic conditions. In itself, then, an increase in T_1 is not pathognomonic for malignant tissue changes.

Table 7 shows the T_1 values from 20 women with confirmed carcinoma of the breast. The data demonstrate of the wide range of individual T_1 values that may occur. The T_1 relaxation times of the neoplastic tissue are increased by an average of about 50 ms. However, the standard deviations show that a positive diagnosis of carcinoma cannot be made on the basis of T_1 values alone, due to the extent of individual variations.

In many pathologic processes T_1 and T_2 are changed in different ways. The T_1 and T_2 values of normal and diseased tissue samples from the human pancreas are

Table 7. T_1 relaxation times of healthy and tumorous breast tissue samples at 0.09 tesla (After Keeler [134])

Patient	T_1 (tumorous) [s]	T_1 (normal) [s]	Difference [s]
1	184	170	14
2	258	171	87
3	200	163	37
4	283	170	113
5	180	146	34
6	210	154	46
7	206	137	69
8	168	109	59
9	164	112	52
10	172	145	27
11	245	188	57
12	181	146	35
13	159	134	25
14	146	99	53
15	250	168	82
16	226	179	47
17	225	139	86
18	216	168	48
19	204	200	4
20	223	173	50
Mean	205	154	50
standard deviation	33	32	23

compared in Fig. 87. It is apparent, for example, that a pseudocyst cannot be distinguished from normal tissue by its T_2 value, but only by its T_1 value. Pancreatitis, on the other hand, is associated with a change in T_2, but not in T_1. For effective tissue discrimination, therefore, it is necessary to obtain tomograms which contain information on both relaxation times.

A fundamental difficulty in the comparative evaluation of NMR tomograms formed with different imaging systems lies in the dependence of the relaxation times on magnetic field strength. While the T_2 values remain largely unchanged, the T_1 values are materially altered by the magnetic field strength (Table 8). Thus, the reporting of T_1 values is meaningful only if the magnetic field strength is also specified.

4.5.2 Pertinent Imaging Parameters

Biologic tissues can be examined for three NMR-relevant properties: water content and the two relaxation times, T_1 and T_2. It is still too early to make a more definite assessment of these quantities in terms of their usefulness for tissue discrimination in NMR images. Experience to date indicates that the most pronounced changes may occur in any of these three parameters, depending on the type of tissue in-

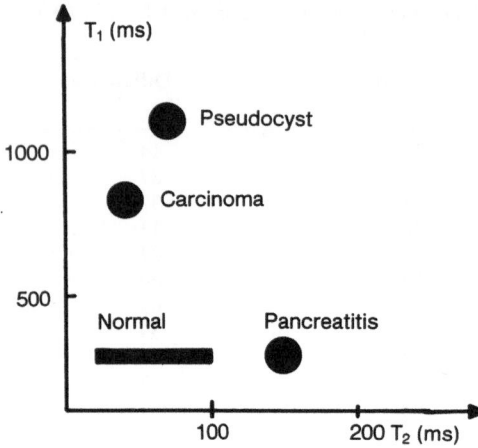

Fig. 87. T_1 and T_2 relaxation times of normal and diseased pancreatic tissue. The values measured in vivo show that some diseases are associated with changes only in the T_1 value (pseudocyst), and some with changes in T_2 (pancreatitis) ($B_o = 0.2$ T). [After Rupp N et al. (1983) Eur J Radiol 3: 68]

Table 8. T_1 relaxation times (msec) of various tissues in relation to magnetic field strength (tesla) [After Rupp N et al. (1983) Eur J Radiol 3: 68]

Tissue	Magnetic field strength		
	0.04 T	0.15 T	0.2 T
Liver	140–170	210	380
Spleen	250–290	–	420
Fat	130–160	–	240
Biliary fluid	400	550–900	890
Ascites	1000	–	2000
Hepatoma	300–450	460–530	570

volved. Therefore, *the goal of NMR tomography must be to obtain images with information on N(H), T_1 and T_2* so that all tissue properties that are accessible to NMR imaging can be included in the medical evaluation. NMR tomograms in which image intensity is determined by water content, T_1, or T_2 can be obtained by appropriate selection of the imaging parameters T_E and T_R. Figure 88 illustrates the importance of these parameters in the time diagram of the signal acquisition process.

Equation (11), which describes image intensity (see Fig. 90), can be separated into three independent factors which define the relationship between tissue parameters and imaging parameters. Thus, the influence of T_2 on the final image can be controlled by varying the delay time T_E, and the influence of T_1 by varying the repetition time T_R. The relationships shown in Table 9 are consequences of Eq. (11).

The inversion recovery sequence (Sect. 4.3.1) can also be used to obtain NMR tomograms that are determined mainly by T_1 (Fig. 89).

The signal intensity of one picture element in the IR image is defined by a modification of Eq. (11):

$$I = N(H) \cdot \exp\left(-\frac{T_E}{T_2}\right) \left[1 - \left\{2 - \exp\left(-\frac{T_R}{T_1}\right)\right\} \exp\left(-\frac{T_I}{T_1}\right)\right] \qquad (14)$$

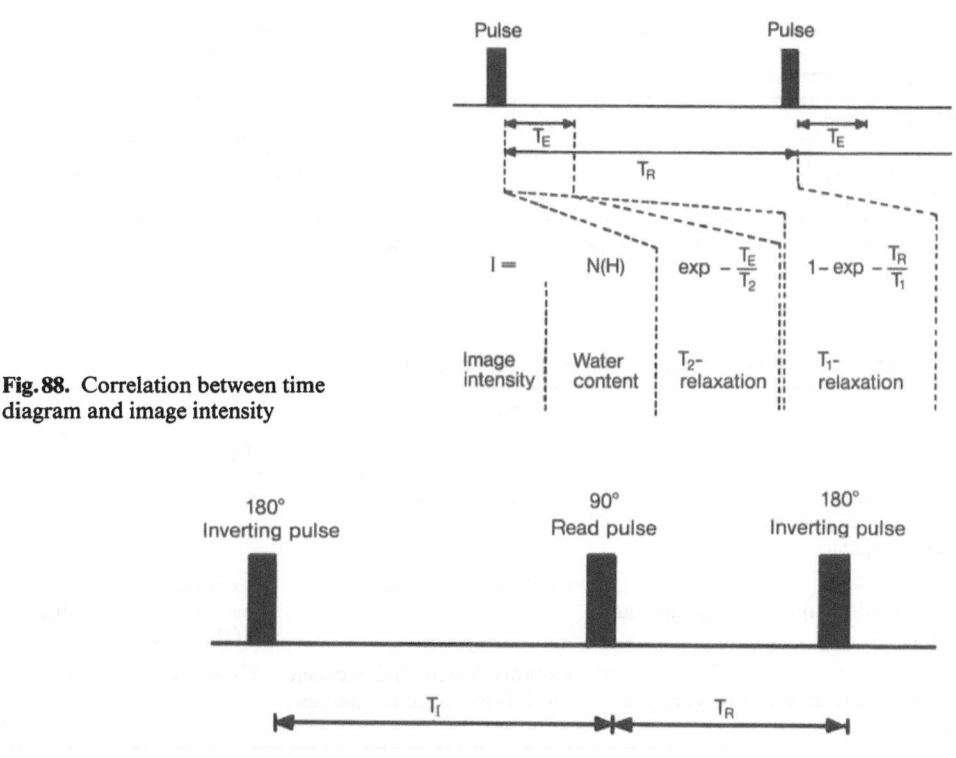

Fig. 88. Correlation between time diagram and image intensity

Fig. 89. Time diagram of the inversion-recovery (IR) technique. After the 180° inverting pulse, the magnetization is pointing in the −z direction. During the interval T_I the magnetization returns to equilibrium in the +z direction. This relaxation process can be detected in the receiver coil by applying a 90° read pulse

For long time delays between signal acquisitions ($T_R \gg T_I$), Eq. (14) is simplified to

$$I = N(H) \cdot \exp\left(-\frac{T_E}{T_2}\right) \left[1 - 2\exp\left(-\frac{T_I}{T_1}\right)\right] \qquad (15)$$

The brightness levels in the tomogram are additionally influenced by the interval T_I between the 180° inverting pulse and the 90° read pulse. The manipulation of T_I, like that of T_R, provides a means of varying the contribution of T_1 to the tomogram without changing the influence of water content. With T_I, however, it is possible to enhance the T_1 information content of the image to a much greater degree than with T_R, and even to cause a reversal of intensity (Figs. 71, 72), thereby expanding the available range of contrasts.

Overall, a highly complex relationship exists among NMR image intensity, the tissue properties N(H), T_1, and T_2, and the imaging parameters T_E, T_R, and T_I [Eq. (11)]. Mathematical analysis reveals that only in the case of a small T_E and large T_R is the NMR tomogram determined by a *single* tissue property – water content. By

Table 9. Effect of imaging parameters on the tissue properties which contribute to the NMR image

Imaging parameters	Tissue properties
Short T_E, long T_R[a]	N(H)
Short T_E, short T_R	N(H), T_1
Long T_E, long T_R	N(H), T_2
Long T_E, short T_R	N(H), T_1, T_2

[a] Large or small relative to the corresponding relaxation time

Table 10. Effect of imaging parameters on image contrast

	Increase of	
	T_E	T_R
Water content: Tissues with high water contents appear *lighter*	Contrast is reduced	Contrast is reduced
T_1: Tissues with large T_1 always appear *darker*	Contrast is reduced	Contrast is maximal between the 2 T_1 values of the tissues
T_2: Tissues with large T_2 always appear *lighter*	Contrast is maximal between the 2 T_2 values of the tissues	Contrast is reduced

and large, image intensity is influenced both by the water content and by at least one of the two relaxation times.

By varying the parameters T_E and T_R (and T_I in the inversion recovery sequence), the relative weight of one of the two relaxation times in the NMR tomogram can be accentuated. NMR tomograms which represent only T_1 or T_2 values cannot be acquired directly, but they can be reconstructed from two tomograms obtained with different sets of imaging parameters.

In conclusion, the diagnostic value of an NMR tomogram relies not on the image intensity itself, but on image contrast, i.e., the signal (brightness) difference between two tissues that must be differentiated. The object, then, of optimizing the parameters T_E and T_R is to translate differences in the tissue properties N(H), T_1 and T_2 into the highest degree of image contrast possible. Table 10 shows clearly that a change in one imaging parameter can lead to a gain or loss of image contrast, depending on the NMR properties of the tissue. This ambiguity is a major problem of practical NMR tomography.

The determination of optimum imaging parameters is complicated by other factors, as well. For example, many pathologic processes are associated with an elevated water content as well as with an increase in both relaxation times. According to Table 4, an increased water content and a prolonged T_2 tend to increase image brightness, while an increase in T_1 tends to reduce it. As a result, both effects may cancel out in certain circumstances, and normal and diseased tissue may not be dif-

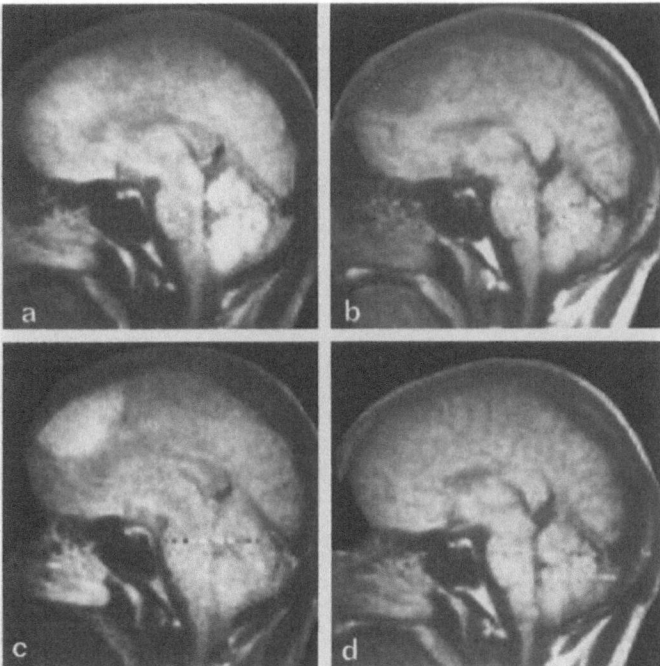

Fig. 90 a–d. NMR tomograms of a grade II astrocytoma. Imaging parameters: **a** $T_R = 1600$ ms; $T_E = 33$ ms. **b** $T_R = 300$ ms; $T_E = 33$ ms. **c** $T_R = 1600$ ms; $T_E = 66$ ms. **d** $T_R = 300$ ms; $T_E = 66$ ms. This series of images clearly documents the effect of operating conditions on the final tomogram. Shortening the repetition time T_R causes the tumor to appear darker in the image (greater T_1), while increasing T_E causes the astrocytoma (longer T_2) to appear lighter. Changing both parameters in the direction indicated gives an NMR tomogram in which the tumor is not demonstrated [From Huk W (1983) In: Wende S, Thelen M (eds) Kernspin-Tomographie in der Medizin. Springer, Berlin Heidelberg New York Tokyo]

ferentiated. A typical example is shown in Fig. 90. Under imaging conditions which essentially yield a water-content tomogram, i.e., a long T_R (1600 ms) and short T_E (33 ms), the tumor, a grade-II astrocytoma, is indistinguishable from surrounding brain tissue (Fig. 90 a). Shortening T_R increases the T_1 information content of the final tomogram, and the tumor, which has a T_1 value greater than that of adjacent tissues, appears *dark* in the image (Fig. 90 b). On the other hand, increasing T_E to 66 ms enhances the T_2 information content of the tomogram, causing the tumor, with its longer spin-spin relaxation time, to appear *light* (Fig. 90 c). The *simultaneous* shortening of T_R and lengthening of T_E causes the opposite effects of T_1 and T_2 on image intensity to cancel, producing a tomogram in which the tumor is not demonstrated (Fig. 90 d).

Because even an ordinary NMR tomographic examination takes several minutes to complete, the systematic variation of imaging conditions requires extra examining time. Thus, a major goal of clinical trials with NMR imagers will be to develop reliable, standard imaging parameters for a wide range of diagnostic situations.

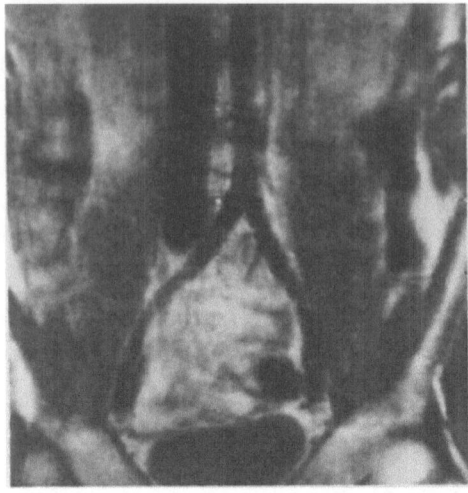

Fig. 91. Coronary NMR tomogram of the abdomen. Operating conditions: imaging time 12 min; $T_R = 400$ ms; $T_E = 39$ ms. Owing to the low image intensities, the vena cava, urinary bladder, and especially the bifurcation of the abdominal aorta are demonstrated in good contrast. (Photo from Siemens, Erlangen)

4.5.3 Effects of Flow

In apparent contradiction to what has already been said, large vessels that contain flowing blood register as dark on the NMR image, despite the fact that blood is more than 98% water (Fig. 91).

The absence of an NMR signal in flowing blood is based on the fact that it takes a finite amount of time to record the signal emitted by the resonating nuclei. This is shown schematically in Figure 92.

After the nuclei have been excited in the selected plane, the actual data acquisition commences after an interval T_E, and the entire data-acquisition process is repeated after a waiting period T_R. In a stationary fluid, the intensity of the signal decays during T_E by T_2 relaxation, and it increases during T_R by T_1 relaxation [see Eq. (11)]. In a slow-moving fluid, there is very little movement of nuclei out of the image plane while the signal is being registered, and the slow flow does not reduce signal intensity during T_E. After the signal has been collected ($T_R = 400–2000$ ms) a certain proportion of the nuclei, which are only partial saturated, leave the region of interest. When the next excitation pulse is applied, therefore, the image plane contains nuclei which were not previously excited and so are still in thermal equilibrium. This creates the appearance of a shorter T_1 relaxation time, leading to an increase in intensity. In fast-moving fluids, on the other hand, the excited nuclei leave the image plane during T_E, and do not generate an NMR signal. *High* flow velocities, then, lead to a *reduction* of image intensity. These relationships between flow velocity and signal intensity are shown graphically in Fig. 93.

The resolution currently available with NMR tomographs permits predominantly major vessels to be imaged. Owing to the high flow velocities in these vessels (0–100 cm/s in the human aorta), they usually appear dark in the image.

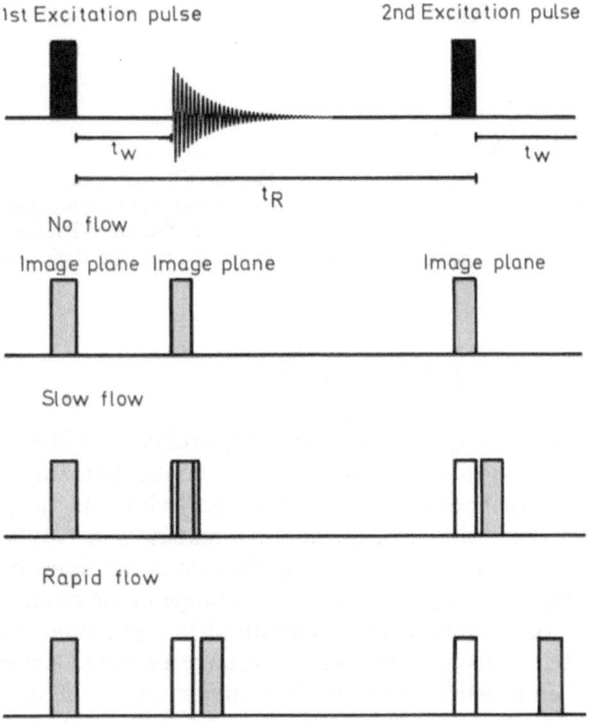

1st Excitation pulse   2nd Excitation pulse

t_W t_W

t_R

No flow

Image plane Image plane   Image plane

Slow flow

Rapid flow

Fig. 92. Time diagram of a flowing fluid. Stationary body fluids give a high-intensity NMR signal owing to their high water content and long relaxation times. In slow-moving fluids, signal intensity is further increased because saturated nuclei will have left the image plane by the start of the next excitation pulse. In fast-moving fluids, the excited nuclei will have already left the image plane during T_E, before the start of actual data acquisition, and no signal is obtained

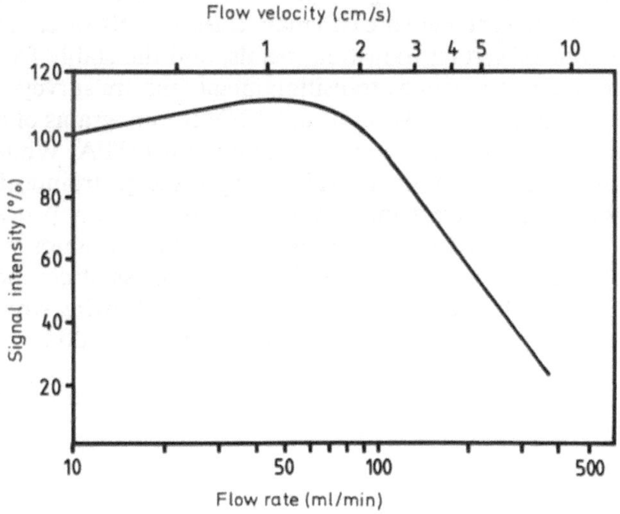

Flow velocity (cm/s)

Fig. 93. NMR signal intensity as a function of flow velocity. As the curve indicates, the relative signal intensity of a fluid with NMR properties comparable to those of blood ($T_1 = 520$ ms, $T_2 = 230$ ms) shows a marked dependence on flow velocity. The fluid was contained in a glass tube with a 9.6-mm inside diameter ($T_R = 500$ ms, $T_E = 43$ ms). [After Crooks L et al. (1982) Radiology 144: 843]

Signal intensity (%)

Flow rate (ml/min)

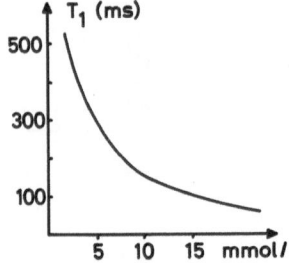

Fig. 94. Change in the T_1 relaxation time of pure water with the concentration of paramagnetic nickel chloride. [After Keeler EK (1983) In: Wende S, Thelen M (eds) Kernspin-Tomographie in der Medizin. Springer, Berlin Heidelberg New York Tokyo]

4.5.4 NMR Contrast Media

Tissue contrast resulting from differences in water content and relaxation times can be enhanced by the use of NMR "contrast media". Unlike radiography, which uses contrast media that are made directly visible by the presence of heavy atoms such as barium and iodine, the contrast materials in NMR tomography can act only indirectly, mainly by shortening the relaxation times of the tissue water. The result of this, according to Eq. (11), is a change in the relative NMR image intensity.

Any paramagnetic compound is a potential NMR contrast medium. This includes ions or molecules which have a strong magnetic moment due to the presence of an unpaired electron. The interaction of the unpaired electron with the spin of the hydrogen nuclei leads to a shortening of the relaxation time of the tissue water, an effect which increases with the concentration of the contrast material (Fig. 94).

At present there are two main classes of NMR contrast media – the paramagnetic ions of certain transition metals, and the stable free radicals. The properties of certain paramagnetic transition metal ions are surveyed in Table 11.

Figures 95 and 96 show axial NMR tomograms of a rat following the oral or intravenous administration of gadolinium DTPA. While orally administered media are better for high-contrast imaging of the gastrointestinal tract, intravenous media provide a clearer demonstration of inflammatory processes and ischemic areas.

The stable free radicals are organic compounds which have a single unpaired electron. Typical representatives of this class of compounds are piperidine-N-oxyl and pyrrolidine-N-oxyl, which have the following structural formulas (where R is a general substituent that is easily varied chemically):

Piperidine-N-oxyl   Pyrrolidine-N-oxyl

Table 11. Magnetic and biologic properties of certain transition metal ions

Ion	Compound(s) Intraperitoneal LD_{50} (mg/kg b. w.)	Tissue affinity after i. v. administration	Magnetic moment[a]
Manganese Mn^{2+}	$MnSO_4 4H_2O$ 534	Liver, kidney	5.9
Iron	$FeCl_3$ 260		5.2
Fe^{2+}, Fe^{3+}	$FeCl_2$ 93		5.9
Copper Cu^{2+}	$CuSO_4$ 5	Liver, blood, bone marrow	1.9
Cobalt Co^{2+}	$CoCl_2 6H_2O$ 90	Thyroid, liver	5.0
Chromium Cr^{3+}	$CrCl_3 6H_2O$ 520	Lung, liver, spleen	3.8
Nickel Ni^{2+}	$NiCl_2$ 26	Brain, lung, heart	3.2
Gadolinium Gd^{3+}	$GdCl_3$ 378	Liver, spleen, muscle	7.9

[a] In units of Bohr magneton

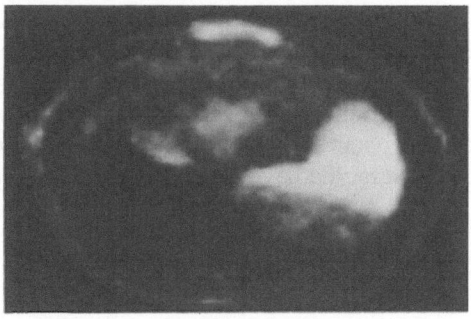

Fig. 95. Axial NMR tomogram of a rat. Thirty minutes after the oral administration of 5 ml of a 1-mmol solution of gadolinium DTPA dimeglumin, the filled stomach of a 300-g rat is demonstrated in very high contrast owing to the shortened T_1 relaxation time of the water. Imaging parameters: $T_E = 28$ ms, $T_R = 500$ ms. (Photo from Schering, Berlin; instrument: 0.35 T system of the University of California at San Francisco)

Brasch et al. (1983) were able to show that the enhancement of image contrast by the administration of stable free radicals is comparable to the effect of paramagnetic metal ions. The functional status of rat kidneys, for example, could be accurately assessed following i. v. administration of stable free radicals.

One advantage of free radicals over metal ions is that the chemical framework into which the radical is incorporated can be easily modified chemically. This makes it possible to introduce the paramagnetic component into a variety of biomolecules.

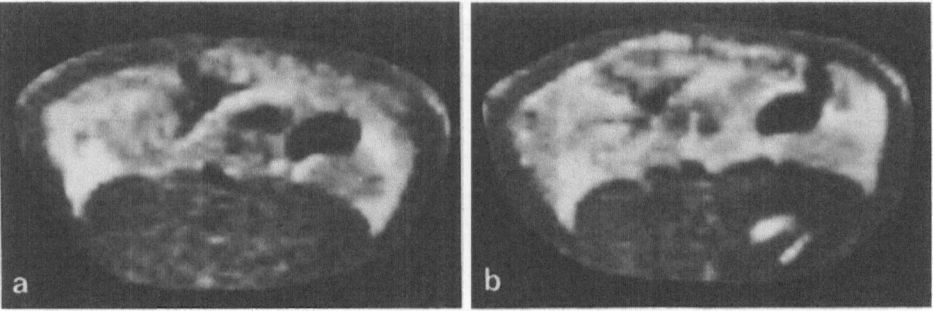

Fig. 96 a, b. Axial NMR tomogram of a rat. **a** The axial NMR tomogram of a 300-g rat does not demonstrate a back-muscle inflammation induced by carrageenan. **b** Ninty minutes after the intravenous administration of 1 mmol/kg b. w. gadolinium DTPA dimeglumin, the inflammation is well delineated. Imaging parameters: $T_E = 28$ ms, $T_R = 500$ ms. (Photo from Schering, Berlin; instrument: 0.35 T system of the University of California at San Francisco)

A great many other paramagnetic compounds are known in addition to those cited above. For example, molecular oxygen is paramagnetic, and, in fact, NMR tomograms of rabbits ventilated with pure oxygen show a heightening of contrast between the blood and myocardium.

Due to a lack of more extensive data, it is not yet possible to make a definitive evaluation of the various NMR contrast media. In particular, animal studies need to be done on the relationship between the concentration of the contrast medium and the observed change of image contrast as a function of the chemical structure of the contrast material. Nevertheless, the ease with which the paramagnetic compounds can be chemically modified promises a wide range of applications once selective contrast media have been developed for specific diagnostic inquiries.

4.5.5 Imaging Time

Due to the small population difference between nuclear magnets in the parallel and antiparallel states, the NMR experiment is relatively insensitive, and adequate image quality can be achieved only by sampling a signal from each element of the image many times, and then summing the data in a computer. The time T needed to collect one free induction signal is given by the formula

$$T = T_E + T_R$$

where T_E is the period between the excitation pulse and the start of data acquisition and T_R the repetition time (see Fig. 90).

Since T_R ranges between 300 and 2000 ms, a total imaging time of 3–15 min is needed to generate a complete NMR tomogram. Use of the multiplanar or Carr-Purcell (multiple spin-echo) technique (see Section 4.4) improves the time efficiency of the imaging process. The time needed to make a *single* tomogram is then given by:

Multiplanar: $T_{total} = (T_E + T_R)/n$

where n is the number of image planes (see Fig. 80),

Carr-Purcell: $T_{total} = (nT_E + T_R)/n$

where n is the number of spin echoes (see Fig. 80).

With these techniques, a *single* tomogram can be generated in about 30 s. The *total* imaging time needed to generate several tomograms is still on the order of several minutes, however.

Another important correlation exists between total imaging time and local resolution. To improve resolution in the image plane by a factor of 2, the number of picture elements must be increased by a factor of 4. To keep the same image quality (signal-to-noise ratio) would require a 16-fold increase in the total imaging time.

4.5.6 NMR Tomography of Elements Other than Hydrogen

Biologic tissues contain a great many elements besides hydrogen. Because all contain at least one NMR-active isotope, it is possible, in principle, to make tomograms based upon other nuclei. The NMR-active nuclei that occur in tissues are listed in Table 12.

Table 12. NMR sensitivity of various nuclei. The total sensitivity takes into account the natural abundance of the NMR-active isotope, its NMR sensitivity, and the occurrence of the element in tissues

Element	Isotope	Total sensitivity
Hydrogen	^{1}H	1.000
Deuterium	^{2}H	$6.2 \cdot 10^{-5}$
Boron	^{11}B	$7.3 \cdot 10^{-7}$
Carbon	^{13}C	$2.5 \cdot 10^{-4}$
Nitrogen	^{14}N	$3.1 \cdot 10^{-3}$
Nitrogen	^{15}N	$6.0 \cdot 10^{-6}$
Oxygen	^{17}O	$4.9 \cdot 10^{-4}$
Fluorine	^{19}F	$6.3 \cdot 10^{-5}$
Sodium	^{23}Na	$1 \cdot 10^{-3}$
Magnesium	^{25}Mg	$7 \cdot 10^{-6}$
Aluminium	^{27}Al	$9 \cdot 10^{-7}$
Silicon	^{29}Si	$9.2 \cdot 10^{-8}$
Phosphorus	^{31}P	$1.4 \cdot 10^{-3}$
Sulfur	^{33}S	$1.1 \cdot 10^{-6}$
Chlorine	^{35}Cl	$8.4 \cdot 10^{-5}$
Potassium	^{39}K	$1.1 \cdot 10^{-4}$
Calcium	^{43}Ca	$9.1 \cdot 10^{-6}$
Iron	^{57}Fe	$5.2 \cdot 10^{-9}$
Copper	^{63}Cu	$8.5 \cdot 10^{-8}$
Zinc	^{67}Zn	$1.8 \cdot 10^{-7}$
Iodine	^{127}J	$2.0 \cdot 10^{-8}$
Lead	^{207}Pb	$3 \cdot 10^{-9}$

The numerical data in Table 12 can give only a rough approximation of NMR sensitivities, because the elemental composition of the organs is not constant (e.g., the iodine content of the thyroid), and because a large portion of the elements are incorporated into high polymers or solids (e.g., fluorine in teeth and bone), and their signal decays so rapidly (short T_2) that it cannot be detected.

Nevertheless, it is clear from the table that the relative NMR sensitivities of these elements are vastly lower than that of hydrogen (by at least 3 orders of magnitude). This suggests the extreme technical difficulties that are involved in the imaging of nuclei other than hydrogen.

In principle, the nuclei of nitrogen (^{14}N), phosphorus (^{31}P), and sodium (^{23}Na) would be the most favorable candidates for NMR imaging from a technical standpoint. However, nitrogen-14 has such an extremely short T_2 relaxation time that it would not be possible to acquire individual NMR spectra from this nucleus, and an image could not be formed.

Using the ^{23}Na nucleus, DeLayre (1981) recorded the NMR tomogram of an isolated, perfused rat heart by adding 145 mmol of the sodium ion to the perfusate. The total imaging time was 20 min at a frequency of 95 MHz.

^{31}P NMR tomograms would be of greatest medical interest, because the distribution of phosphorus metabolites reflects the energy metabolism of tissues. As detailed in the first part of the book, conventional NMR spectra are composed of several ^{31}P signals which are assigned to the phosphorus metabolites ATP, phosphocreatine, and inorganic phosphate. Thus, the individual projections that are obtained after the application of a magnetic field gradient are based upon multiple individual signals. Lauterbur has developed mathematical formulae for generating NMR tomograms based on each of the phosphorus metabolites (Fig. 97).

In biologic tissues, of course, the total measurable phosphorus concentration is considerably lower (50 mmol/l) than in the phantom used in Fig. 97. However, as the whole-body imaging must be done at a lower frequency due to the skin effect (see Section 2.1), ^{31}P NMR tomography in living subjects may be impractible.

Three other NMR-active nuclei are of potential importance for NMR imaging; these are carbon-13, deuterium, and fluorine. All three nuclei have a very low abundance in tissues, but their levels can be increased by the administration of suitable compounds. By recording the corresponding tomograms, the distribution of these "tracer" compounds within the body could be mapped and followed. It should be added, however, that a high concentration of the tracer would have to be present in the tissue due to the inherently low sensitivity of the imaging method.

The use of ^{13}C-labeled compounds is impractical for reasons of cost (about US $ 500/gram of ^{13}C-labeled glucose). On the other hand, heavy water (D_2O) and especially fluorinated blood plasma substitutes could have future importance for imaging of the cardiovascular system.

In summary, it appears that NMR tomography will continue to be based on hydrogen for the forseeable future. Expanding the technique to other nuclei in natural abundance is not feasible due to their extremely low NMR sensitivity, which is at least 1000 times less than that of hydrogen.

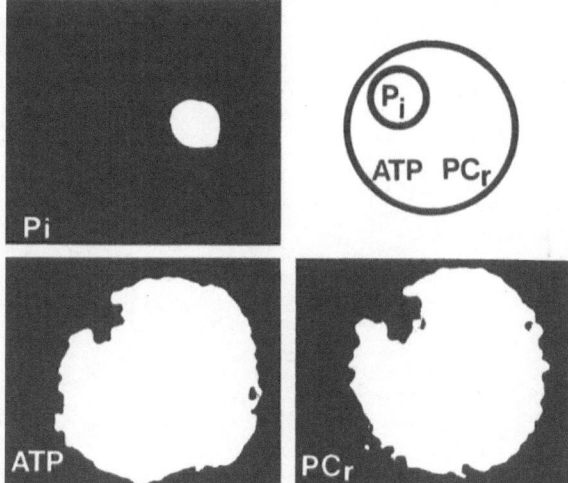

Fig. 97. ^{31}P NMR tomograms of a phantom. The phantom consists of a 15-mm tube filled with an aqueous solution of 0.15 mol ATP and 0.7 mol phosphocreatine, and a 5-mm tube filled with 1.6 mol inorganic phosphate solution (pH = 5.72). Imaging frequency was 146 MHz. Thirty separate projections were taken, and 1024 spectra were accumulated for each projection. The NMR tomograms of the individual metabolites were reconstructed from the individual projections. [From Lauterbur PC et al. (1980) J Magn Reson 38: 343]

4.6 Examples of the First Applications of NMR Tomography

The main goal of this introduction is to present the basic principles of NMR spectroscopy and tomography in terms that are easily understood. A few examples will serve to illustrate the capabilities of modern NMR imaging systems. Because developments in this field are rapid, it is hoped that the examples presented will, above all, convey an idea of the potential of the NMR technique. For readers who wish to consult the original literature, a short bibliography is provided at the end of the chapter.

4.6.1 Head

Because the head is free of intrinsic motion, it is an ideal object for NMR investigations. Even with imaging times of several minutes, very high-quality tomograms of the head can be obtained (Fig. 98).

Axial and sagittal NMR tomograms permit a clear differentiation of many structures including the pituitary, choroidal plexus, cerebellum, and 4th ventricle. The various gyri and sulci of the brain are well delineated. Above all, the gray and white matter of the brain are clearly differentiated. Cavities filled with cerebrospinal fluid generally appear dark; because the excited nuclei in the highly mobile fluid travel a significant distance during the imaging process they do not yield a detectable NMR signal (see Sect. 4.6.3).

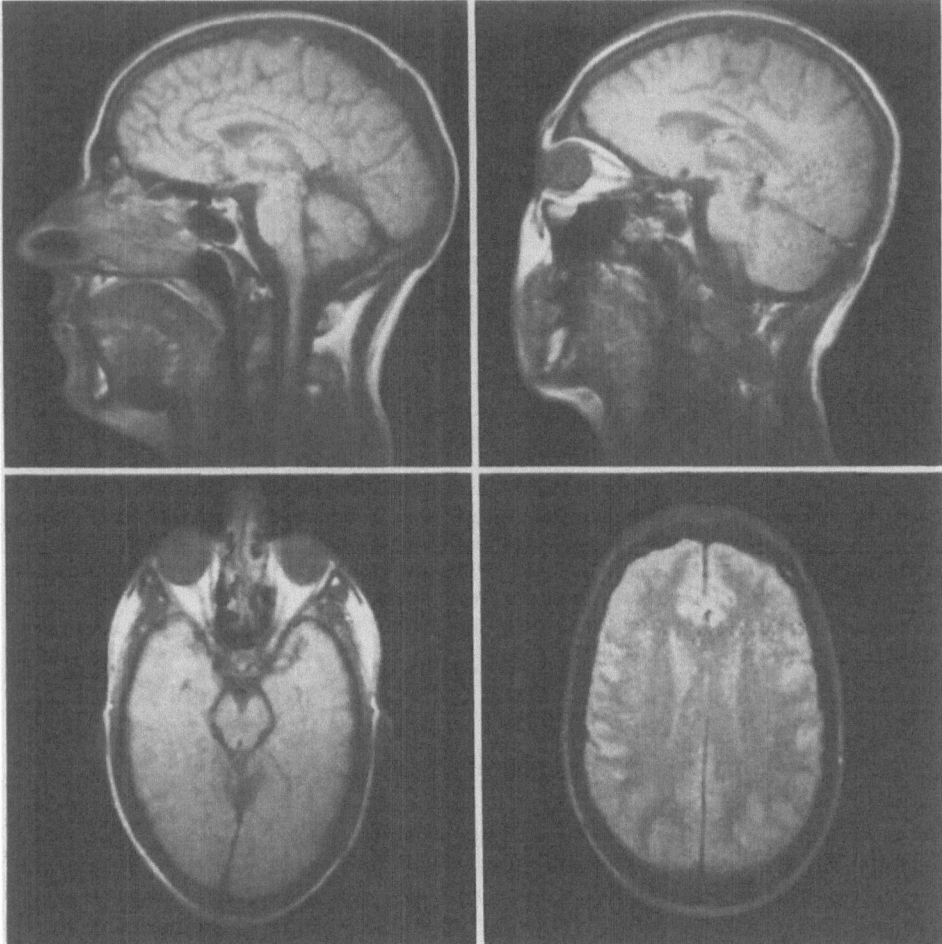

Fig. 98. Axial and sagittal NMR tomograms of the human head. All the tomograms are rich in anatomic detail. Note that the optic nerve, rectus lateralis and rectus medialis muscles are clearly differentiated from adjacent fatty tissue. The basal cisterns, filled with cerebrospinal fluid, appear dark in the image. (Photo from Siemens, Erlangen)

Studies indicate that brain tumors which are demonstrated by CT scans generally are visible on NMR tomograms. A comparison of the abilities of CT and NMR to differentiate a glioblastoma (glioblastoma multiforme) reveals a fundamental difference between the two imaging methods (Fig. 99). In the NMR image, the perifocal edema accompanying the tumor cannot be clearly distinguished from the tumor itself, because the tumor and edema have similar relaxation times and the same high water content.

Figure 100 shows the tomograms of a cerebral abscess. The small abscess and surrounding edema are clearly seen in the CT image. In the NMR proton-density image, on the other hand, the abscess is not differentiated from the inflammatory swelling because the water content of each is increased to an equal degree.

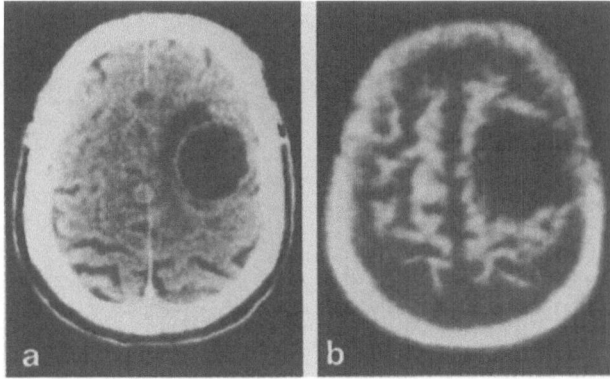

Fig. 99 a, b. CT scan (**a**) and NMR tomogram (**b**) of a patient with glioblastoma multiforme. A mass is demonstrated in both tomograms. In the NMR image, however, the tumor cannot be distinguished from surrounding edema. The image was made with the inversion recovery technique, causing the tumor, with its long T_1 value, to register as dark. Imaging parameters: inversion recovery sequence ($T_1 = 400$ ms, $T_R = 1000$ ms) with image reconstruction from 180 separate projections using a 256×256 image matrix; section thickness 9 mm; imaging time 4 min. (Photo from Steiner RE, Young IR, London)

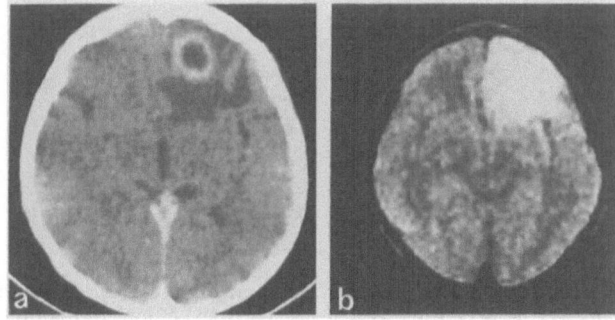

Fig. 100 a, b. CT scan (**a**) and NMR tomogram (**b**) of a cerebral abscess. Imaging parameters: $T_R = 1720$ ms, $T_E = 36$ ms. While the CT scan demonstrates the ring structure of the abscess in the surrounding edema, the tumor cannot be distinguished from the edema in the NMR tomogram, because both the abscess and the inflammatory edema have the same, increased water content. (Photo from Dr. W. Huk, University of Erlangen)

NMR generally permits an earlier diagnosis of cerebral infarction than the CT scan does because the infarction causes an immediate lengthening of the T_1 relaxation time in the affected tissue. This leads to a marked decrease in image intensity that is most apparent in the inversion recovery image (Fig. 101).

Based on experience to date, NMR tomography is excellently suited for diagnosing the various stages of multiple sclerosis. The prolonged relaxation time of the plaque causes the focal lesions to stand out in high contrast and enables their size and distribution to be easily determined (Fig. 102).

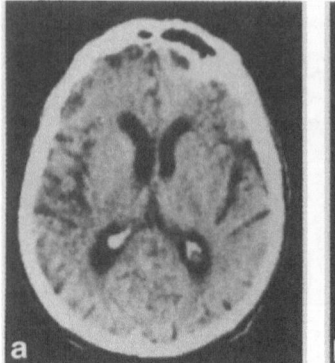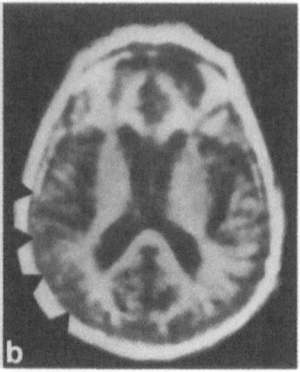

Fig. 101 a, b. CT scan (**a**) and NMR tomogram (**b**) of a cerebral infarction. Imaging conditions same as in Fig. 99. The CT scan demonstrates a weakly attenuating lesion and a widening of the sulci in the area of the left posterior operculum. The inversion-recovery NMR tomogram demonstrates a considerably larger lesion of the operculum *(arrows)*. (Photo from Steiner Re, Young IR et al., Hammersmith Hospital, London)

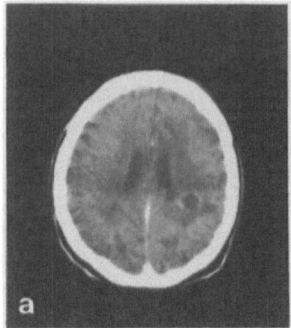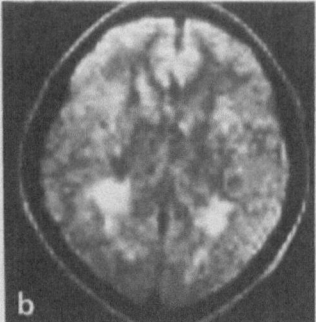

Fig. 102 a, b. CT scan (**a**) and NMR tomogram (**b**) of a patient with multiple sclerosis. Imaging parameters: $T_R = 1720$ ms, $T_E = 35$ ms. The focal lesions of multiple sclerosis are demonstrated much more clearly in the NMR image than in the CT scan. Under the imaging conditions indicated, the increase in relaxation time caused by the demyelinating process causes relatively intense signal generation by the plaque. [From Huk W (1983) In: Wende S, Thelen M (eds) Kernspin-Tomographie in der Medizin. Springer, Berlin Heidelberg New York Tokyo]

4.6.2 Torso

NMR tomography of the spine gives a clear picture of the intervertebral discs. The ease of sagittal-plane imaging with NMR represents a major advantage over CT for the diagnosis of disease in this region (Fig. 103).

Because imaging takes several minutes to complete, involuntary movements of the organs in the thoracic cavity and retroperitoneum associated with breathing, heart function, and peristalsis may lead to motion artifacts. Despite these limitations, however, it has been found that very high quality NMR tomograms of the thoracic and abdominal organs can be obtained. The diagnostic capabilities of

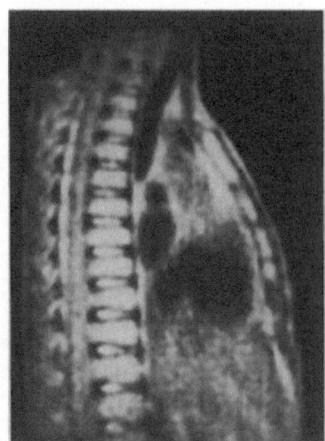

Fig. 103. Mediosagittal NMR tomogram of the thorax. While the organs in the thoracic cavity appear unsharp due to their intrinsic motion, the spine and intervertebral discs are clearly demonstrated. The cord is also well delineated. (Photo from Bruker, Karlsruhe)

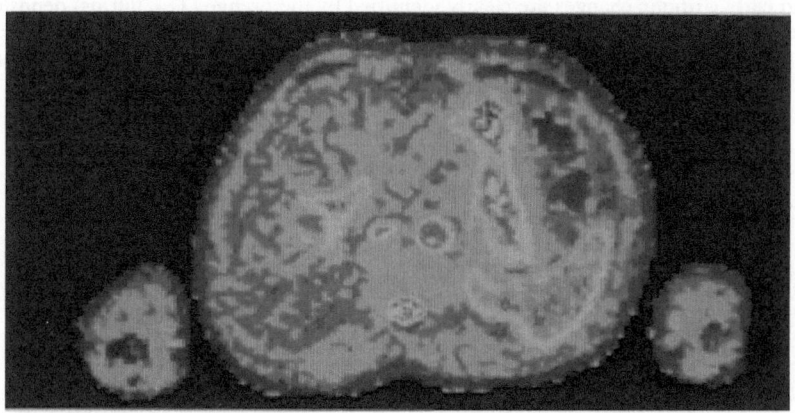

Fig. 104. Axial NMR tomogram of the abdomen. This T_1 image gives a pictorial representation of the T_1 relaxation times of the tissues. Here a color format is used in which short relaxation times register as dark blue, and longer T_1 values register as yellow or red. In this tomogram from a normal control subject, the liver has a T_1 value of 150–155 ms. The porta hepatis, spleen, stomach, abdominal aorta, inferior vena cava and spinal cord are easily identified ($B_o = 0.04$ T). (Photo from Prof. J. R. Mallard, M&D Technology, Aberdeen)

NMR tomography are illustrated by the images of normal and diseased livers presented in Figs. 104–112.

The quality of images in the thoracic cavity can be greatly improved by synchronizing the sequence of NMR imaging with the ECG ("gating"). Under these circumstances, artifact-free images of the heart itself can be obtained (Fig. 113).

Vessels register as dark owing to the presence of flowing blood and generally can be clearly differentiated from surrounding tissues (Fig. 114). The excellent contrast of blood vessels makes it easy to recognize aneurysms and deposits (Figs. 115 and 116).

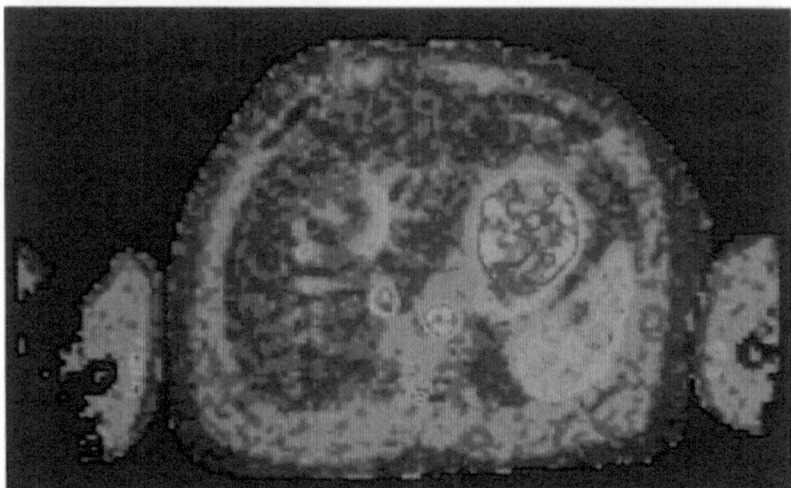

Fig. 105. Alcohol-induced cirrhosis. The fatty infiltration of the liver secondary to alcohol abuse and early cirrhotic changes are clearly identified by the orange ($T_1 = 180$ ms) deposits in the liver, which appears dark blue ($T_1 = 130$–140 ms). The stomach shows heavy mucosal infoldings, and the spleen is slightly enlarged ($B_o = 0.04$ T). (Photo from Prof. J. R. Mallard, M&D Technology, Aberdeen)

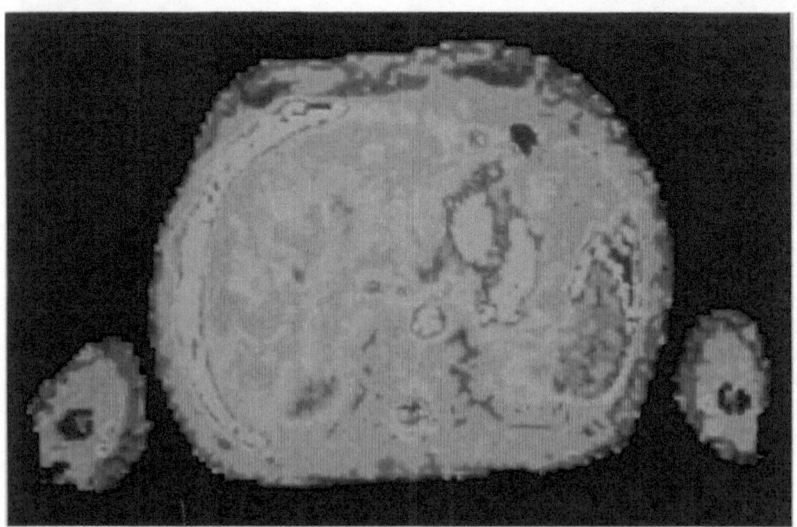

Fig. 106. Cirrhosis with ascites. The T_1 value of the entire liver is markedly increased (180–190 ms). The ascites registers as white due to its long T_1 value of 650–700 ms. The enlarged spleen is also well demonstrated ($B_o = 0.04$ T). (Photo from Prof. J. R. Mallard, M&D Technology, Aberdeen)

Fig. 109. Hepatoma. The NMR tomogram demonstrates the hepatoma in the left half of the liver ▶ ($T_1 = 340$–350 ms). The spleen is markedly enlarged and has displaced the stomach ($B_o = 0.04$ T). (Photo from Prof. J. R. Mallard, M&D Technology, Aberdeen)

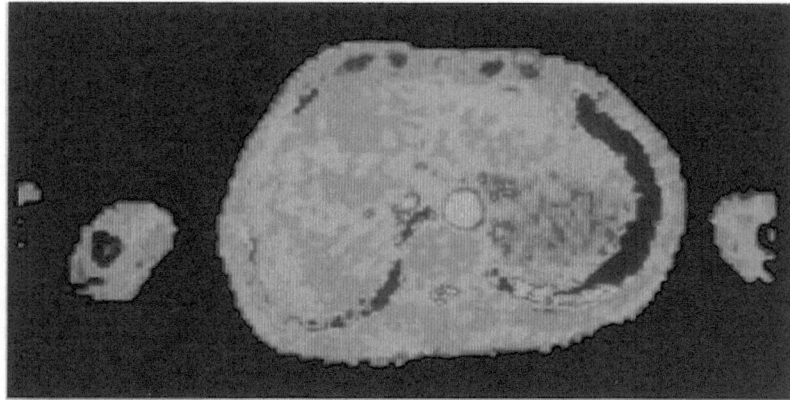

Fig. 107. Chronic hepatitis. The liver shows a markedly prolonged T_1 relaxation time of 180–190 ms. The tomogram shows a portion of the basal lung regions *(black)* and a fluid collection in the pleural region *(white)*. (Photo from Prof. J. R. Mallard, M&D Technology, Aberdeen)

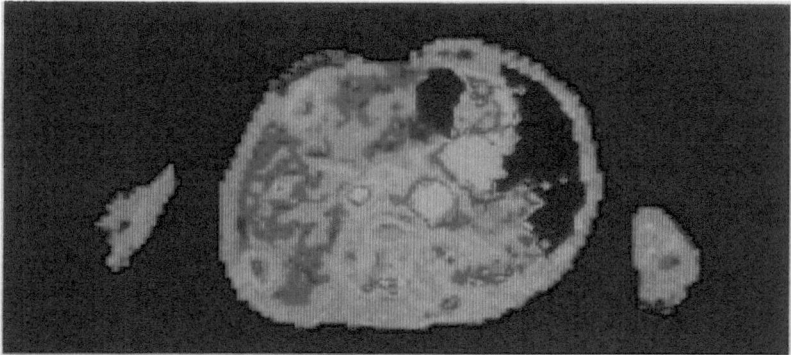

Fig. 108. Metastases. Circular metastases ($T_1 = 350$–360 ms) from a primary carcinoma of the sigmoid colon. The T_1 value of the healthy liver tissue is normal ($T_1 = 160$ ms). Air in the bowel *(black)* and fluid in the stomach *(white)* are clearly visible, as are the inferior vena cava and abdominal aorta ($B_o = 0.04$ T). (Photo from Prof. J. R. Mallard, M&D Technology, Aberdeen)

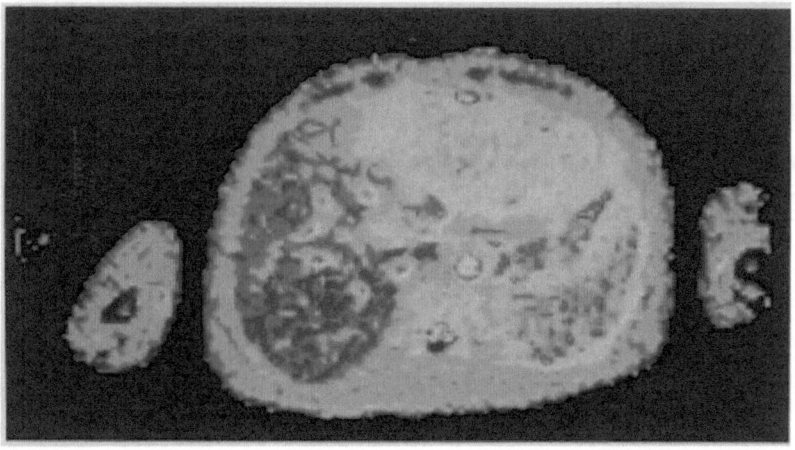

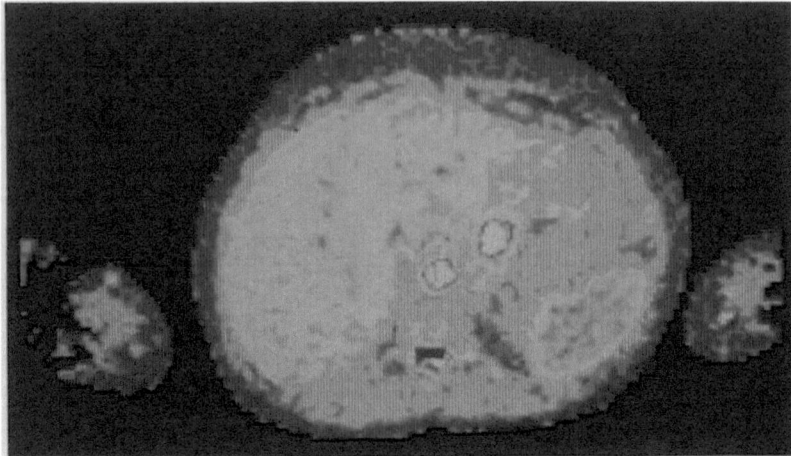

Fig. 110. Cholangiocarcinoma. The tumor, with a T_1 value of 350 ms, involves practically the entire liver. The T_1 value of the tumor is similar to that occurring in cirrhosis, but the more uniform distribution serves to distinguish the tumor from that condition ($B_o = 0.04$ T). (Photo from Prof. J. R. Mallard, M&D Technology, Aberdeen)

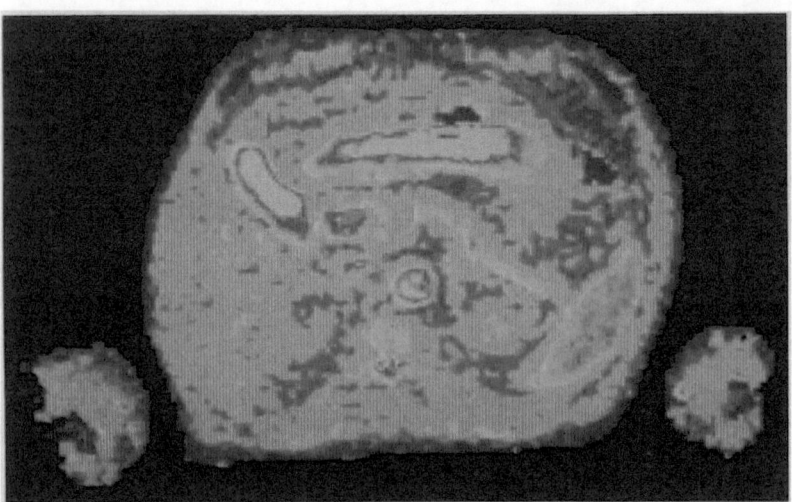

Fig. 111. Obstructive jaundice. The gallbladder stands out clearly in the NMR tomogram due to obstruction of the common bile duct by edema following the passage of a stone. The splenic vein is also demonstrated ($B_o = 0.04$ T). (Photo from Prof. J. R. Mallard, M&D Technology, Aberdeen)

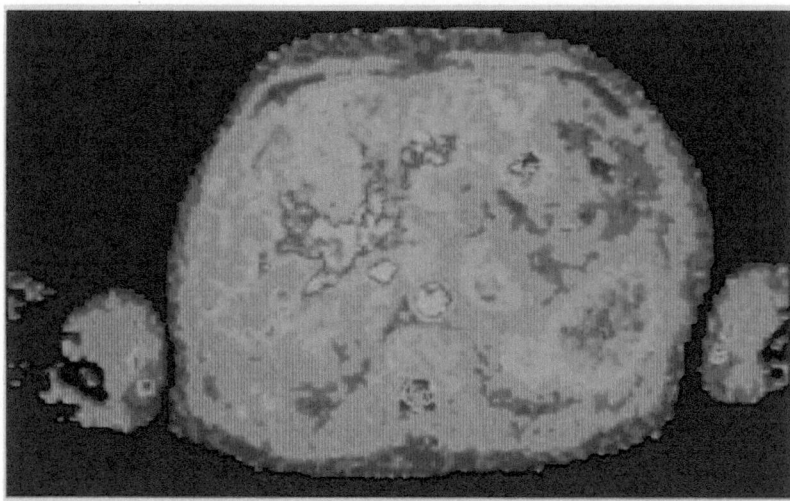

Fig. 112. Obstructive jaundice associated with pancreatic carcinoma. The carcinoma has led to congestion of the common bile duct and intrahepatic biliary tract. The intrahepatic ducts are characterized by a prolonged T_1 and display a typical pattern ($B_o = 0.04$ T). (Photo from Prof. J. R. Mallard, M&D Technology, Aberdeen)

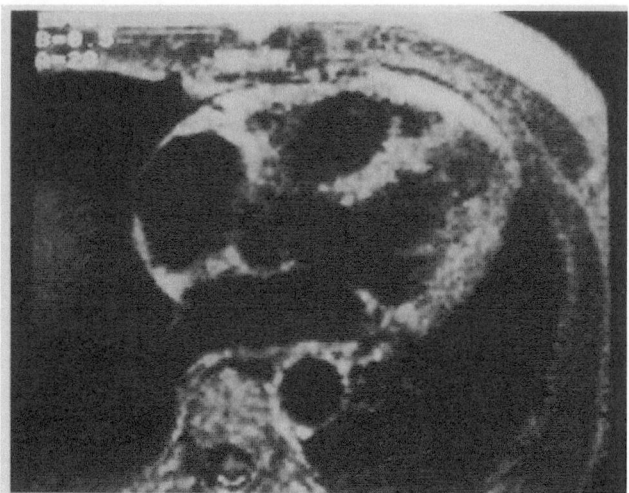

Fig. 113. Axial NMR tomogram of the heart. The intervals of NMR signal acquisition are synchronized with the ECG to obtain artifact-free images of the heart. The atria, ventricles, interatrial and interventricular septum, and the aorta are clearly demonstrated. (Photo from Prof. L. Kaufman, Diasonics, San Francisco) [From Schaaf H (1983) In: Wende S, Thelen M (eds) Kernspin-Tomographie in der Medizin. Springer, Berlin Heidelberg New York Tokyo]

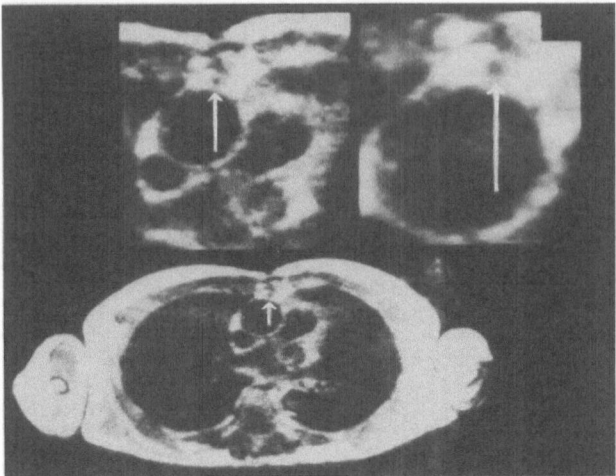

Fig. 114. Axial NMR tomogram of the thoracic cavity. Imaging parameters: $T_R = 500$ ms, $T_E = 28$ ms. The identification of a coronary bypass illustrates the high resolving power of NMR tomography. (Photo from Prof. L. Kaufman, Diasonics, San Francisco) [From Schaaf H (1983) In: Wende S, Thelen M (eds), Kernspin-Tomographie in der Medizin. Springer, Berlin Heidelberg New York Tokyo]

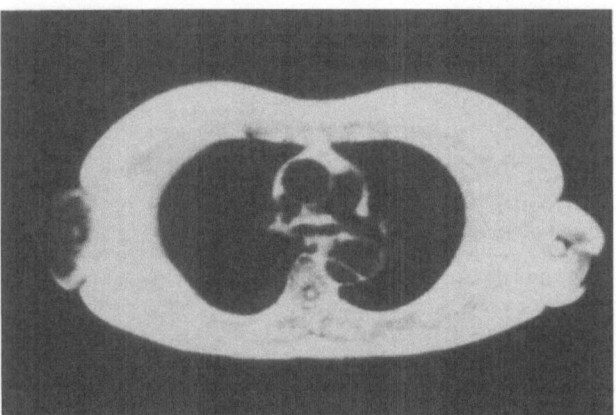

Fig. 115. NMR tomogram of a dissecting aortic aneurysm. Imaging parameters: $T_R = 1000$ ms, $T_E = 28$ ms. The dissection is clearly seen. The aortic lumen and the area of dissection appear dark due to the presence of flowing blood. Anteriorally the ascending aorta, pulmonary artery, superior vena cava and tracheal bifurcation are well demonstrated. (Photo from Prof. L. Kaufman, Diasonics, San Francisco) [From Schaaf H (1983) In: Wende S, Thelen M (eds) Kernspin-Tomographie in der Medizin. Springer, Berlin Heidelberg New York Tokyo]

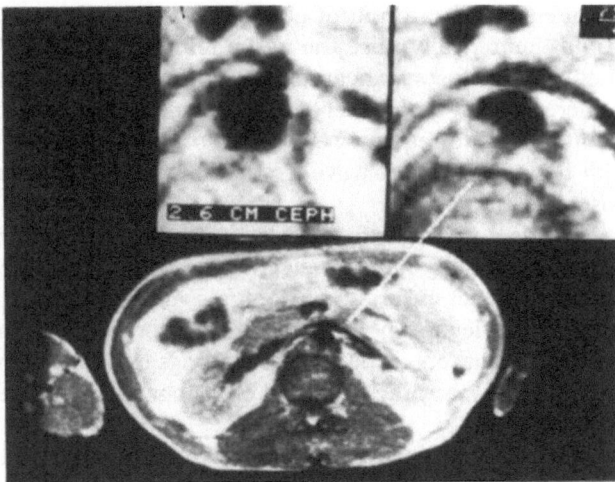

Fig. 116. Axial NMR tomogram of the abdomen. Imaging parameters: $T_R = 1000$ ms, $T_E = 28$ ms. The inset *(top right)* clearly shows a plaque in the abdominal aorta, which is not visible 2.6 cm farther caudally *(top center)*. Vascular anatomy is well demonstrated. Note the mesenteric artery and vein, the left renal vein, and the crura of the diaphragm. [From Schaaf H (1983) In: Wende S, Thelen M (eds) Kernspin-Tomographie in der Medizin. Springer, Berlin Heidelberg New York Tokyo]

4.7 Health Risks of NMR Tomography

4.7.1 Physiologic Effects of Magnetic and Radiofrequency Fields

During the several minutes needed to produce an NMR tomogram the human body is exposed to: the static, homogeneous magnetic field, the changing magnetic fields used to form and switch the magnetic field gradients, and the irradiating radiofrequency energy. Below we shall examine briefly the question of possible health risks associated with exposure to these fields.

The effect of *static magnetic fields* on biologic systems has long been an object of experimental study, but a confusing picture emerges when published results are reviewed. Initially a marked increase in T-wave amplitude of the ECG of laboratory animals exposed to magnetic fields stronger than 0.3 tesla was reported. Later, an exhaustive study was done in which this phenomenon was attributed solely to the induction of a voltage by blood flowing perpendicular to the magnetic field. No blood pressure changes, arrhythmias, or changes in heart rate were observed. It appears, then, that the rise of T-wave amplitude has no physiologic causes or effects and persists only as long as the field is switched on.

Genetic studies in *Drosophila melanogaster* exposed to magnetic fields of 3.7 T showed no demonstrable genetic injury. Data published previously on changes in enzyme activities could not be reproduced in fields of up to 20 T.

The physiologic effect of the *radiofrequency radiation* is based on the heating of tissues by the absorbed energy. Under the conditions of NMR spectroscopy, a maximum energy absorption of about 4 W/kg is observed. Over a 10-min period this would theoretically raise the tissue temperature by 0.7 °C. However, this assumes that the tissue loses *no* heat to the environment through evaporation or radiation. It is therefore likely that very little actual tissue heating will take place. The total amount of radiofrequency energy absorbed is substantially less than that associated with therapeutic short-wave irradiation, for example, and is extremely unlikely to pose a health risk.

The *time-varying magnetic fields* used in NMR tomography are capable of inducing potentially harmful electric currents within the body. There have been reports of changes in certain physiologic values such as urinary hormone levels in rats and a fall of leukocyte counts in mice exposed to changing magnetic fields. However, many authors dispute these findings as evidence of adverse health effects, because a great many factors could cause changes in these extremely labile physiologic parameters.

According to a model calculation by Budinger (1981), a magnetic field changing at a rate of 1 T/s induces a current density of 1 μA/cm^2. By comparison, the current density associated with the action potential of a nerve is about 3000 μA/cm^2. Since current densities above 300 μA/cm^2 can provoke ventricular fibrillation, this effect may be the greatest health hazard of NMR tomography. It should be emphasized, however, that under the operating conditions ordinarily used for NMR imaging, the current densities that occur are at least 2 orders of magnitude below that critical level.

One well-documented effect of a changing magnetic field on the human body is the occurrence of "magnetic phosphenes". These are flickers of light and swirls of color which are perceived within the eye upon exposure to a magnetic field of over 10 mT alternating at a frequency of 20–40 Hz. These visual disturbances, which appear to be completely reversible, are probably due to the effect of the magnetic field on the light-sensitive photoreceptors of the retina.

Only recently have systematic studies been conducted on the biologic effects of NMR tomography in its entirety, i. e., circumstances in which all energy fields are simultaneously active. Exhaustive studies in bacteria and lymphocytes have demonstrated no injurious effects.

In summary, there is strong evidence that the static and time-varying magnetic fields and radiofrequency fields used in NMR tomography pose a substantially lower health risk than the ionizing radiation used in X-ray CT, the physiologic effects of which have long been known. But while no irreversible harm has yet been demonstrated, the growing importance of NMR in medicine makes it imperative that further research be done before the method is declared to be completely safe.

4.7.2 Risks from Metallic Implants

Besides the effects of NMR tomography on tissues, consideration must be given to its effects on metallic objects implanted within the body, such as cardiac pacemakers and surgical clips.

Modern cardiac pacemakers (weight between 50 and 70 g) are usually mounted in a metal casing made of stainless steel or titanium. Virtually all pacemakers currently in use are of the "demand" type. Here, the electrical signals from the device (typical values: 5 V, 10 mA, and 1 ms) are applied directly to the myocardium via special sensing circuitry whenever the heart rate falls below a critical value. Pacemaker malfunctions may have a variety of causes. Given the wide range of pacemaker designs presently on the market, it is not possible to make a blanket statement concerning their susceptibility to malfunction under the conditions of NMR imaging. However, the major causes of pacemaker failure under such circumstances can be discussed.

In many pacemaker models the static magnetic field closes a relay, thereby inactivating the sensing circuitry which monitors heart activity, and causing the pacemaker to generate pacing impulses whether they are needed or not. A magnetic field strength of only 0.1 mT is sufficient to cause this effect, and a field of this intensity may be experienced at distances of 5 m or more from a 0.5 T magnet. This is not considered dangerous, however, because the same effect, produced by holding a small magnet against the patient's chest, is used routinely as a clinical test of pacemaker function.

With some pacemakers, a far more serious danger is posed by the radiofrequency radiation and time-varying magnetic field gradients, which can mimic normal cardiac activity so that pacing impulses are not supplied when needed. In addition, pacemakers may move within the tissue of the chest wall when placed in a magnetic field. At present, patients with cardiac pacemakers should not be subjected to NMR tomography and should keep at least 6 m away from the center of the magnet.

Metallic objects implanted within the body may lead to complications during NMR imaging. Nonferromagnetic materials such as various amalgams and gold do not cause artifacts, but ferromagnetic metals such as those occurring in certain steel dental appliances can seriously impair image quality in tomograms of the head. Of particular importance is the effect of the magnetic field on permanent surgical implants, such as hemostatic clips that contain ferromagnetic materials. Forces may be exerted on these implants that are sufficient to cause displacement or dislodgement, with a substantial degree of risk to the patient. However, it is reasonable to assume that as NMR gains wider acceptance in human medicine, increasing use will be made of nonferromagnetic alloys and metals for the manufacture of surgical clips.

4.7.3 Safety Recommendations

This chapter concludes with a list of NMR safety standards that have been advocated by certain organizations.

1. *National Radiological Protection Board, United Kingdom (1980)*
   Static magnetic fields: <2.5 T
   Switched gradient fields: <20 T/s
   Radiofrequency radiation: up to 15 MHz and <1 W/kg

2. *National Radiation Protection Board (United States)*
   Static magnetic fields: <2.5 T
   Switched gradient fields: <25 T/s
   Radiofrequency radiation: <70 W (whole body)

References for Chap. 4

Monographs and Progress Reports

NMR Imaging in biomedicine. Mansfield P, Morris PG, Academic Press, London New York, 1982
Nuclear magnetic resonance imaging in medicine. Kaufman L, Crooks LE, Margulis AR, Igakun-Shoin, Tokyo, 1981
Kernspin-Tomographie in der Medizin. Wende S, Thelen M (Hrsg), Springer, Berlin Heidelberg New York Tokyo, 1983

The First NMR-Tomogram

Image formation by induced local interactions: Examples employing nuclear magnetic resonance. Lauterbur PC, Nature 242: 190 (1973)

Survey Articles

NMR-Spektroskopie am Menschen. Limbach HH, Nach Chem Techn Lab 28: 860 (1980)
Principles and methods of imaging by proton NMR. Chambron J, Armspach JP, Wecker D, J Biophys Med Nucl 5: 89 (1981)
Physical Principles of NMR-tomography. Loeffler W, Oppelt A, Eur J Radiol 1: 338 (1981)
Kernspin-Tomographie. Ganssen A et al., Computer-Tomogr 10 (1981)
Die Kernspintomographie (KST) und ihre klinischen Anwendungsmöglichkeiten. Zeitler E, Schittenhelm R, Electromedica 49: 2 (1981)
NMR-Tomographie. Roth K, Gronenborn A, Chem i u Zeit 16: 35 (1982)
Kernspinresonanz-Tomographie. Habermehl A, Graul EH, Dtsch Aerztebl 79: 17 (1982)
Initial clinical evaluation of a whole body NMR tomograph. Young IR et al., J Comput Assist Tomogr 6: 1 (1982)
NMR imaging in medicine. Pykett IL, Sci Am 246 (5): 54 (1982)
Principles of NMR imaging. Pykett IL et al., Radiology 143: 157 (1982)
Bildgebende Kernresonanz. Stetter E, Kastler J, Funkschau 1982: 43
Kernspin-Tomographie: Bilder aus torkelnden Atomkernen. Karcher HL, Selecta 50: 4674 (1982)
NMR imaging techniques and applications: A review. Bottomley PA, Rev Sci Instrum 53: 1319 (1982)
NMR imaging. Andrew ER, Acc Chem Res 16: 114 (1983)
Kernspin-Tomographie – "Röntgen" ohne Strahlenbelastung. Zeitler E et al., Dtsch Apothek Z 123: 241 (1983)
Magnetische Kernresonanz. Strecker E, Dtsch Med Wochenschr 108: 551 (1983)
NMR-Tomographie. Buchmann F, Heinzerling J, GIT Lab Med 6: 102 (1983)
Kernmagnetische Resonanz in der Medizin. Oppelt A, Physik i u Zeit 14: 7 (1983)
The diagnostic value of morphology and relaxations time in NMR-imaging of the body. Rupp N, Reiser M, Stetter E, Eur J Radiol 3: 68 (1983)

Imaging Parameters

Visualization of cerebral and vascular abnormalities by NMR imaging. The effect of imaging parameters on contrast. Crooks LE et al., Radiology 144: 843 (1982)
Clinical efficiency of NMR imaging. Crooks LE et al., Radiology 146: 123 (1983)
Signal, noise, and contrast in NMR imaging. Edelstein WA et al., J Comput Assist Tomogr 7: 391 (1983)

Selected Original Publications

Brain

NMR tomography of the brain. Holland GN et al., J Comput Assist Tomogr 4: 1 (1980)
NMR tomography of the brain: Coronal and sagittal sections. Holland GN et al., J Comput Assist Tomogr 4: 429 (1980)

NMR tomography of the brain: a preliminary clinical assessment with demonstration of pathology. Hawkes RC et al., J Comput Assist Tomogr 4: 577 (1980)

Imaging of the brain by NMR. Doyle FH et al., Lancet II: 53 (1981)

NMR observations in alcoholic cerebral disorder and the role of vasopressin. Besson JAO et al., Lancet II: 923 (1981)

NMR imaging of brain tumours unrevealed by CT. Einsiedel H Gräfin von, Löffler W, Eur J Radiol 2: 226 (1982)

NMR imaging in Wilson disease. Steiner RE, Young IR et al., J Comput Assist Tomogr 7: 1 (1983)

NMR imaging of Arnold-Chiari type I malformation with hydromyelia. Buonanno FS et al., J Comput Assist Tomogr 7: 126 (1983)

NMR-imaging in white matter disease of the brain using spin-echo sequences. Young IR et al., J Comput Assist Tomogr 7: 290 (1983)

NMR imaging of the brain in systemic lupus erythematosus. Steiner RE et al., J Comput Assist Tomogr 7: 461 (1983)

NMR tomography of the central nervous system: Comparison of two imaging sequences. Huk W et al., J Comput Assist Tomogr 7: 468 (1983)

NMR-Untersuchungen bei Erkrankungen des Gehirns und Rückenmarkes. Huk W. In: Kernspin-Tomographie in der Medizin. Wende S, Thelen M (Hrsg) Springer, Berlin Heidelberg New York Tokyo, 1983

Vergleich von NMR und CT anhand direkter Sagittal-Schnitte des Gehirnschädels. Blümm RG. In: Kernspin-Tomographie in der Medizin. Wende S, Thelen M (Hrsg) Springer, Berlin Heidelberg New York Tokyo, 1983

Cardiovascular System

NMR tomography of the normal heart. Hawkes RC et al., J Comput Assist Tomogr 5: 605 (1981)

NMR imaging of the cardiovascular system: Normal and pathologic findings. Herfkens RJ et al., Radiology 147: 749 (1983)

NMR imaging of the infarcted muscle: A rat model. Kaufman L et al., Radiology 147: 761 (1983)

Three-dimensional display of NMR cardiovascular images. Bottomley PA et al., J Comput Assist Tomogr 7: 172 (1983)

NMR imaging of atherosclerotic disease. Herfkens RJ et al., Radiology 148: 161 (1983)

Erste Ergebnisse der Kernspin-Tomographie bei Gefäßerkrankungen. Zeitler E et al. In: Kernspin-Tomographie in der Medizin. Wende S, Thelen M (Hrsg) Springer, Berlin Heidelberg New York Tokyo, 1983

Thorax and Breast

Oesophageal carcinoma demonstrated by whole-body NMR imaging. Smith FW et al., Br Med J 282: 510 (1981)

NMR imaging and evaluation of human breast tissue: Preliminary clinical trials. Ross RJ et al., Radiology 143: 195 (1982)

NMR imaging of the thorax. Gamsu G et al., Radiology 147: 473 (1983)

Initial experience with NMR imaging of the human breast. El Yousef SA et al., J Comput Assist Tomogr 7: 215 (1983)

Clinical application of NMR using FONAR technique in diseases of the breast and lung. Keeler EK. In: Kernspin-Tomographie in der Medizin. Wende S, Thelen M (Hrsg) Springer, Berlin Heidelberg New York Tokyo, 1983

Abdomen and Retroperitoneum

NMR tomography of the normal abdomen. Hawkes RC, J Comput Assist Tomogr 5: 613 (1981)

NMR imaging of the liver: Initial experience, NMR tomographic imaging in liver disease. Smith FW et al., Lancet I: 963: (1981)

NMR imaging of the pancreas. Smith FW et al., Radiology 142: 677 (1982)

NMR imaging of the kidney. Crooks LE, Kaufman L et al., Radiology 146: 425 (1983), 147: 765 (1983)

NMR imaging of the gallbladder. Crooks LE, Kaufman L et al., Radiology 147: 481 (1983)

NMR imaging of the adrenal gland: A preliminary report. Crooks LE et al., Radiology 147: 155 (1983)

NMR imaging of induced renal lesions. London DA, Radiology 148: 167 (1983)

Die Kernspin-Tomographie des Abdomens und des Beckens. Rödl W, Lutz H, Oppelt A. In: Kernspin-Tomographie in der Medizin. Wende S, Thelen M (Hrsg) Springer, Berlin Heidelberg New York Tokyo, 1983

Ear, Nose, and Media

Work in progress: NMR anatomy of the larynx and tongue base. Lufkin RB et al., Radiology 148: 173 (1983)

NMR: Normale und pathologische Befunde im HNO Bereich. Zeitler E et al. In: Kernspin-Tomographie in der Medizin. Wende S, Thelen M (Hrsg) Springer, Berlin Heidelberg New York Tokyo, 1983

NMR Contrast Media

Relaxation Rate Enhancement observed in vivo by NMR imaging. Doyle FH et al., J Comput Assist Tomogr 5: 295 (1981)

Ansatzmöglichkeiten für Kontrastmittelanwendungen in der Kernspin-Tomographie. Niendorf HP, Weinmann HJ. In: Kernspin-Tomographie in der Medizin. Wende S, Thelen M (Hrsg) Springer, Berlin Heidelberg New York Tokyo, 1983

NMR study of a paramagnetic nitroxide contrast agent for enhancement of renal structures in experimental animals. Brasch RC et al., Radiology 147: 773 (1983)

Potential oral and intravenous paramagnetic NMR contrast agents. Runge VM et al., Radiology 147: 789 (1983)

Methods of contrast enhancement for NMR imaging and potential applications. Brasch RC, Radiology 147: 781 (1983)

Health Risks of NMR

Magnetic field effects on biological systems. Tenford TS, Plenum Press, New York, 1979

NMR in vivo Studies: Known thresholds for health effects. Budinger TF, J Comput Assist Tomogr 5: 800 (1981)

The effects of NMR exposure on living organisms I: A microbial assay. Thomas A, Morris PG, Br J Radiol 54: 615 (1981)

The effects of NMR exposure on living organisms: A genetic study of human lymphocytes. Cooke P, Morris PG, Br J Radiol 54: 622 (1981)

The effects of NMR on patients with cardiac pacemakers. Pavlicek W et al., Radiology 147: 149 (1983)

Potential hazards and artifacts of ferromagnetic and nonferromagnetic surgical and dental materials and devices in NMR imaging. New PFJ et al., Radiology 147: 139 (1983)

Risiken und Gefahren der NMR-Tomographie. Rinck PA, Dtsch Med Wochenschr 108: 992 (1983)

5 Outlook

Even the earliest applications of NMR spectroscopy and tomography to biochemical and medical inquiries proved the importance and versatility of this new diagnostic technique. Potential applications range from the detection of enzyme-deficiency diseases by the assay of phosphorus metabolites to the diagnosis of neoplasms from NMR tomograms. It would be premature to make a definitive comparison between NMR and other, established modalities, because the routine clinical use of whole-body NMR systems on a broad basis has been possible for only a short time. Also, it is difficult to appreciate the full potential of NMR at the present time, because the correlation between pertinent tissue properties and the operating parameters of the NMR system has not yet been fully elucidated. Already, however, the basic advantages and disadvantages of the NMR technique are becoming apparent.

Advantages of NMR
- Completely noninvasive
- Appears to have no ill effects

NMR spectroscopy:
- Enables the noninvasive, biochemical examination of otherwise inaccessible organs, such as the brain, by the use of surface coils.

NMR tomography:
- Enables soft-tissue imaging
- Image contrast can be altered by manipulating instrument parameters

Disadvantages of NMR
- High cost of whole-body NMR systems
- High installation costs in many instances
- Not recommended for patients with cardiac pacemakers or metallic implants

NMR spectroscopy:
- Relatively high magnetic field strengths are needed, requiring the use of costly superconducting magnets

NMR tomography:
- Relatively slow imaging time (several minutes)
- Poorer resolution than with X-ray CT (but better than with scintigraphy)

The advantages and disadvantages listed above clearly show that while NMR cannot totally replace established imaging methods, it is definitely a valuable adjunct to them. We are on the threshold of systematic clinical trials which should answer unresolved questions about the future of NMR spectroscopy and tomography in diagnostic medicine.

Appendix A

The NMR Experiment in the Rotating Coordinate System

For didactic reasons, the discussion of the NMR experiment presented in Sections 2.1 and 4.3 was greatly simplified and does not provide a sufficient basis for exploring the original literature. To equip the reader with facts needed to pursue the subject further, we shall examine the NMR experiment in more detail, introducing the notion of a rotating coordinate system in an effort to deepen our understanding of the technique.

From a quantum-mechanical standpoint, nuclear spins do not align precisely in a parallel and antiparallel fashion when exposed to an external magnetic field. Instead, they assume a certain angle relative to the direction of the applied field (Fig. 117). Thus, the terms "parallel" and "antiparallel" do not accurately describe the spatial orientation of the nuclei.

According to the laws of classical physics, the angle between the nuclear spin and magnetic field axis leads to a rotation, or precession, or the nuclear spin around the magnetic field direction. This is analogous to the precession of a child's top in the earth's gravitational field (Fig. 118).

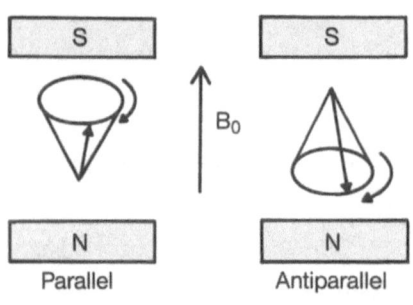

Parallel   Antiparallel

Fig. 117. The precession of nuclear spin about the magnetic field axis

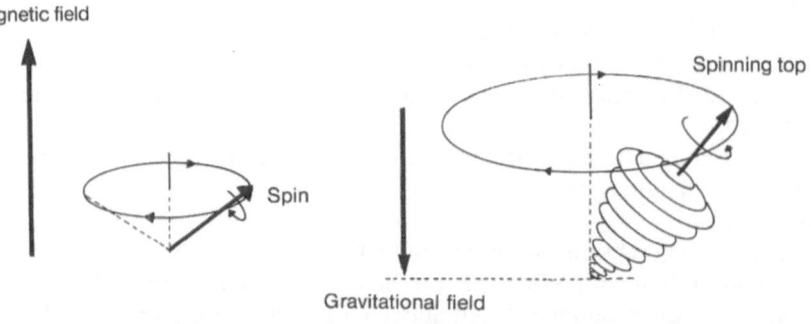

Nuclar spin   Top

Fig. 118. The precession of nuclear spin about the magnetic field axis is analogous to that of a child's top about the direction of the earth's gravitational field

The frequency at which the nuclei precess is given by the Larmor relation

$$v_0 = \frac{\gamma}{2\pi} \cdot B_0 \qquad (16)$$

Because the nuclear magnets sum to give the sample a net magnetization, the relative position of the individual magnets on the cone of precession is immaterial. Through averaging over many nuclei, the net magnetization is aligned with the magnetic field direction and precesses at the frequency v_0 (Fig. 119).

The transmitter coil produces the alternating field, B_1, by applying a radiofrequency pulse perpendicular to the magnetic field direction (x axis). This B_1 field can be divided into two components, or "partial waves," which have opposite directions of rotation ("circumferentially polarized partial waves," see Fig. 120). Only the partial waves whose direction of rotation coincides with the precessional direction of the net magnetization can interact with the nuclei, and only these waves will be considered below.

At resonance, the rotational frequency of the B_1 partial wave is equal to the precessional frequency of the nuclei ($v_1 = v_0$). The rotation of the B_1 partial wave and of the net magnetization makes it more difficult to describe the physical processes associated with NMR in schematic and mathematical terms. It is expedient, therefore, to move from a normal, stationary frame of reference to one which rotates

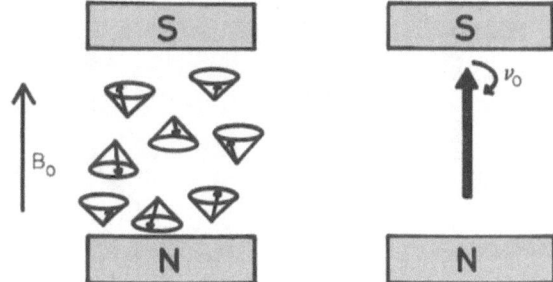

Fig. 119. The summation of nuclear spins leads to a net magnetization in the sample

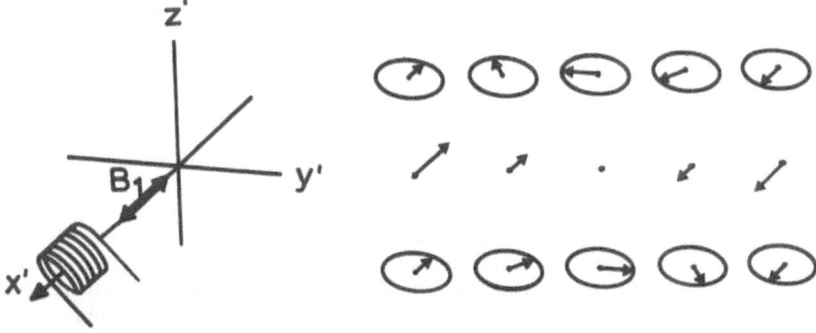

Fig. 120. Separation of the alternating magnetic field B_1 into two, opposite partial waves

about the z axis at the Larmor frequency (*rotating coordinate system*, Fig. 121). The axes of the rotating coordinate system are designated x', y', and z' to distinguish them from the x, y and z axes of the stationary coordinate system. The drawings in Fig. 122 are intended to clarify the notion of a changing frame of reference, which is necessary for a clear understanding of the NMR experiment.

When resonance occurs in the rotating coordinate system, the partial wave of the B_1 field is oriented in the x' direction, and the net magnetization is oriented in the z' direction. In this case M_0 will precess about the B_1 axis in the z'–y' plane (see Fig. 69). Strictly speaking, then, the terms "90° pulse" and "180° pulse" are defined only in a rotating reference frame. After the B_1 pulse is turned off, the magnetization gradually returns to equilibrium by the two relaxation mechanisms (Fig. 70).

While a mathematical description of the NMR experiment is very simple within the context of the rotating coordinate system, the electronic detection of resonance itself can occur only in a stationary receiving coil oriented in the x' direction, so an inverse transformation of the coordinate system would be necessary before a useful signal is acquired. However, through special electronic processing of the original signal (phase-sensitive detection) in the NMR spectrometer, a signal is acquired and stored in the computer which corresponds precisely to conditions in the rotating coordinate system, so an inverse transformation is no longer needed to describe the *processed* signal. In other words, the signal acquired after electronic processing in

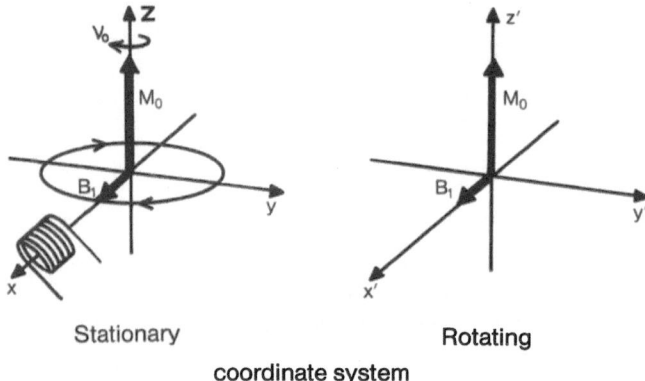

Stationary   Rotating
coordinate system

Fig. 121. A comparison of the stationary and rotating coordinate systems

---

Fig. 122 a, b. The change from a stationary to a rotating coordinate system. Many physical pro- ▶
cesses can be simply described only if a suitable frame of reference is chosen. In some cases, therefore, it is better to dispense with a familiar, static frame of reference in favor of a coordinate system which may at first seem complex. This principle is illustrated by a string duet. **a** The violinist, who is in a rotating frame of reference, is unable to join in with the second musicians because the sheet music is in a different reference frame. On the other hand, the cellist (and we) are able to read the music, because he (and we) are in the stationary reference frame. **b** If the music stand (and we) were to join the violinist in the rotating reference frame (coordinate transformation), then he (and we) could read the music without difficulty. This time it is the cellist, in the stationary reference frame, who is unable to join in

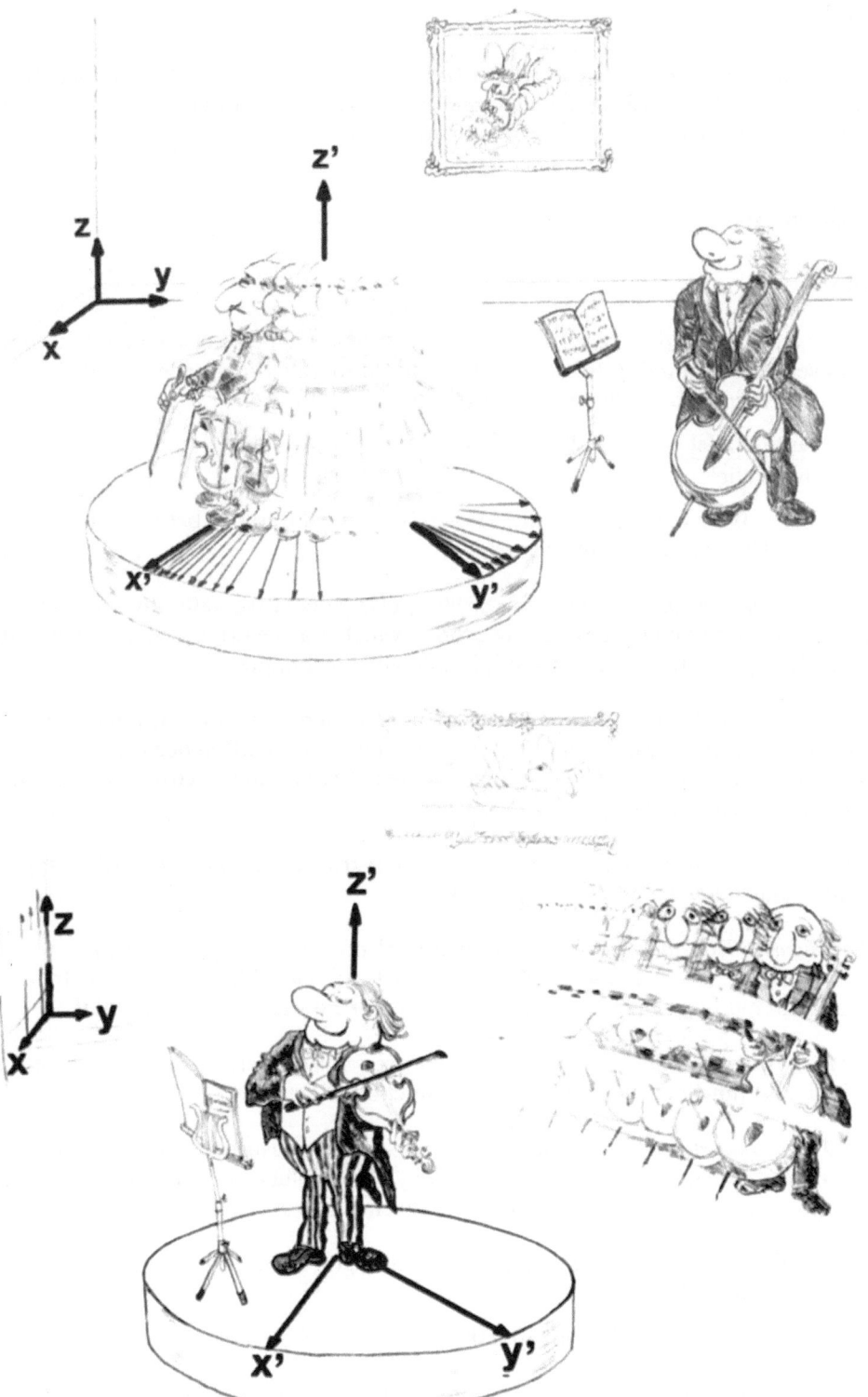

the spectrometer *corresponds to physical processes occurring in the rotating coordinate system*. It is natural and expedient, therefore, to analyze all physical processes within the framework of the rotating coordinate system.

Appendix B

Glossary

Boltzmann Distribution. The distribution of nuclear magnets between the two energy levels, corresponding to the parallel and antiparallel alignments, is described quantitatively by the Boltzmann distribution.

$$\frac{N\,\text{antiparallel}}{N\,\text{parallel}} = \exp(-\frac{\Delta E}{kT}),$$

Where k is Boltzmann's constant, ΔE is the energy difference between the energy levels, and T is absolute temperature.

Carr-Purcell Experiment. Following a 90° pulse, successive 180° inverting pulses are applied to produce multiple spin echoes, which may be processed collectively or individually to obtain images (multiple spin-echo technique).

Chemical Shift. Dependence of the resonance frequency of a nucleus on the chemical binding of the atom or structure of the molecule. This difference is based essentially on an attenuation of the applied magnetic field by the electrons surrounding the nucleus (see *Shielding*).

Cryomagnet. A magnet in which the coil is constructed of an alloy which is superconducting at low temperatures ($<8°$ K).

Fast Fourier Transformation. A computational process, developed by Cooley and Tukey, for the Fourier transformation of digitized data. The process is extremely rapid and has become the standard technique for Fourier analysis in NMR.

Fourier Transformation. A mathematical process for converting a quantity that is a function of time into a quantity that is a function of frequency.

FID (Free Induction Decay). The decay curve that is measured after the application of an excitation pulse. It represents the decay of the voltage induced in the radiofrequency coil by the excitation pulse in the absence of external influences (free).

Gauss. Obsolete measure of magnetic field strength. It has been generally replaced by the ISU unit, the tesla (T).

1 T = 10 000 gauss = 10 kilogauss (kG)

Gradient. A change, usually linear, of magnetic field strength along one spatial direction.

Magnetogyric Ratio. A proportionality constant (γ) between magnetic field strength and resonance frequency. It is unique for each nucleus.

Larmor Frequency. The precessional frequency ν_0 of nuclear magnets perpendicular to the direction of the external magnetic field

$$\nu_0 = \frac{\gamma}{2\pi} B_0$$

where B_0 is the magnetic field strength and γ is the magnetogyric ratio.

Lorentzian Line. A spectral signal of the form

$$I(\nu) = \frac{a}{b + (\nu - \nu_0)^2}$$

is called a Lorentzian line. It is characteristic of NMR signals evoked in an homogeneous magnetic field.

Nuclear Quadrupolar Moment. An electric moment that is caused by the asymmetric distribution of electric charge in the nucleus and is found only in nuclei with a spin number greater than ½. The NMR signals of these nuclei (e. g., ^{14}N) are usually very broad and cannot be used to produce an image.

Nuclear Spin. The intrinsic rotation of nuclei that is responsible for the magnetic properties exploited in the NMR experiment.

$\pi/2$ Pulse, π Pulse. Synonyms for 90° pulse and 180° pulse respectively.

Precession. The rotation of a magnetic moment around the magnetic field axis.

Quadrupolar Nuclei. Nuclei which have a nuclear quadrupolar moment (q. v.).

Quenching. A sudden loss of superconductivity in a cryomagnetic coil due, for example, to a rise of temperature. The total energy of the magnet is suddenly released in the form of heat, causing evaporation of the helium coolant.

Relaxation Times T_1 and T_2. Characteristic time constants which describe the time required to establish a thermal equilibrium of nuclear populations, and which depend on the mobility of the nuclei in the molecule.

Saturation. A condition in which both NMR energy levels are equally populated.

Saturation recovery technique. One 90° pulse or a series of excitation pulses are applied to populate all energy levels equally (saturation). The subsequent relaxation process (recovery) can be observed with a 90° "read" pulse.

Shielding. Attenuation of the external magnetic field at the nucleus of an atom by the opposite field induced by electrons surrounding the nucleus (see *Chemical shift*).

Shim Coils. Correcting coils used to neutralize existing nonhomogeneities of magnetic field strength.

Signal-to-noise Ratio. The relative distance between the signal height and the average noise. For a specified standard solution, the S/N ratio is a measure of the quality of the NMR spectrometer or imager.

Skin Effect. The increase of electrical resistance at high frequencies with increasing depth of penetration.

Spin. The intrinsic rotation of elementary particles (protons, electrons) and several nuclei that is responsible for their magnetic properties.

Spin Decoupling. The spin-spin coupling that is observed in certain cases (e.g., ^{13}C–^1H) can be suppressed by applying a second radiofrequency field (double resonance) in the resonant range of one of the two nuclei.

Spin Density. The number of nuclear magnets per unit volume.

Spin Echo. By applying a 180° inverting pulse one can compensate for the decrease in transverse magnetization caused by magnetic field nonhomogeneities. A spin echo is evoked by the following pulse sequence: $90°$-T_E-$180°$-T_E-echo.

Spin Quantum Number. A characteristic constant for each nucleus; it may be either zero or a multiple of ½. From the spin quantum number (I) can be determined the number of energy states that are possible in an applied magnetic field $(2I+1)$.

Spin-lattice Relaxation Time T_1. The relaxation time T_1 is the characteristic time constant of the growth and decay of longitudinal magnetization. It is determined by the interaction of the nuclear magnets with their environment (lattice).

Spin-spin Coupling. The interaction between NMR-active nuclei within a molecule can lend extra detail to the evoked NMR signals. High resolution is needed to demonstrate this splitting of the signals, however.

Spin-spin Relaxation Time T_2. The T_2 relaxation time is the characteristic time constant of the growth and decay of transverse magnetization. It is determined by the interaction of the nuclear magnets with one another and with their environment.

Superconducting Magnet. See Cryomagnet.

Tesla. The ISU unit of magnetic field strength, T.

Thermal Equilibrium. A distribution among several energy levels of thermal equilibrium corresponding to the Boltzmann distribution.

Topical Magnetic Resonance (TMR) Spectroscopy. Term for the acquisition of an NMR spectrum from a circumscribed region within a larger (biologic) object.

Zeeman Effect. Splitting of the energy levels in an external magnetic field.

Zeugmatography. Term coined by Lauterbur to describe the coupling of the main magnetic field to the radiofrequency field by the nuclear magnets in the object.

6 Addendum

In the short time since the appearance of the German edition, the steadily increasing number of publications on the biological and medical applications of NMR have attested to the efficacy of this technique. The following is designed to provide a brief insight into current research trends. An appended selection from the literature will help facilitate the reader's access to recent developments.

In Vivo NMR Spectroscopy

^{31}P-NMR spectroscopy as a method of investigating cellular metabolism has been applied to a number of problems. Because of technical limitations, many studies concentrated on the investigation of isolated, perfused organs of smaller laboratory animals. By changing the physiological parameters or by introducing drugs during perfusion, their influence on high-energy phosphates can be followed continuously. In this manner Seymour et al. were able to demonstrate that in rat hearts treated with insulin after 18 min of total ischemia, the intracellular pH value sank to 6.0 and the ATP content to 70% of its preischemic level, while untreated hearts under the same conditions showed only 40% ATP and a pH value of 6.3. Reperfusion allowed only the insulin-treated hearts to be restored to a functional condition, leading to the conclusion that irreversible tissue damage is caused by a reduction in ATP level and not by diminished intracellular pH values.

The saturation transfer technique has proved valuable for measuring reaction rates. Matthews et al. were able to show that although temperature reductions in isolated perfused rat hearts left phosphocreatine and ATP concentrations unchanged, the rate of phosphocreatine (PC) transformation to ATP sank drastically.

The brain is especially interesting in animal studies of organs in situ since a surface coil makes it easily accessible to NMR spectroscopy and allows it to be selectively studied in view of the limited number of other noninvasive techniques. Golczewski et al. were able to show that under normal conditions the PC/ATP ratio in rat brains is independent of age. However, under ischemia, it does not change in mature young rats whereas in older animals the ischemia PC/ATP ratio increases significantly, indicating a reduction in ATP synthesis.

Investigations of other atomic nuclei have been conducted in addition to the numerous ^{31}P-NMR studies. Fluorinated anesthetics were observed noninvasively in the brain of intact rabbits with fluorine-19 nmr spectroscopy. Spectra were obtained with a surface coil centered over the calvarium. Residual fluorine-19 signals from halothane could be detected as long as 98 h after termination of anesthesia.

The practicality of carbon-13 nmr for measurement of metabolic flux parameters in living system were demonstrated by adding carbon-13 labelled acetate or pyruvate to perfused rat hearts. The hearts were frozen different times after addition and extracted with perchloric acid. The carbon-13 fractional enrichment of individual carbons of different metabolites were calculated from the area of the resolved resonances at 90 MHz. From these data a well determined citric acid cycle flux of 8.3 µmol/g/min was calculated for an observed oxygen consumption of 31 µmol/g/min.

Whole-body instruments for measuring NMR spectra have already been installed and the first ^{1}H, ^{31}P, and ^{13}C NMR spectra of live adult humans were published recently. These spectra were obtained with a system that allows measurement of both NMR tomograms as well as NMR spectra. This makes possible simultaneous observations of the anatomy and the biochemistry of healthy and diseased tissues.

NMR Tomography

The quality of NMR tomograms has been improved by instrument manufacturers and has in many instances attained a level comparable to that of CT. It has been possible to increase resolution to the submillimeter level. The question of the optimal strength of the magnetic field is for the most part still open. On the one hand, an increase in the magnetic field leads to an improvement in measurement sensitivity; on the other hand, the relaxation times are also increased, so that improved signal-background noise relation is partially sacrificed to longer waiting periods between experiments. Earlier predictions, based on model calculations, that whole-body scans in excess of 20 MHz are impossible due to the limited penetrability of radio frequencies have been proved wrong through impressive abdomen scans at 1.5 T (= 63 MHz) and head scans at 2.0 T (= 87 MHz).

One way of drastically reducing measuring times was already described some time ago by Mansfield. With clever measuring techniques he was able to circumvent complicated reconstructions, making quasi-real time imaging possible. The extremely short scanning time of only 35 ms especially in thorax imaging reduces motion artefacts to zero. Whether developing this method further will result in better resolution remains to be seen.

In the field of NMR contrast media, certain intravenously applied gadolinium compounds proved to be well tolerated; certain intracranial tumors would also be imaged much more clearly with the use of contrast media.

The continuing series of systematic applications of NMR tomography to diagnostic problems in recent months, the reports of which can be found in a large number of publications, has proved again and again the incredible efficacy of this method.

Chemical Shift Imaging

By using high field strengths and pulse sequences with suitable switched gradients, it is possible to combine NMR spectroscopy with tomography. The object of this combination is to produce tomograms whose brightness reflects the concentration of certain metabolites. In initial studies it was possible to arrive at selective tomographic representations of lipid and water distribution in tissues. In addition, one remarkable study was able to produce ^{31}P-NMR tomograms of the ATP, PC, and P_i content of a cat brain in vivo at 2.9 T. Because of the limited natual sensitivity of ^{31}P nuclei, the measuring time was 4 h (resolution: 5 mm). Whether it will be possible to improve resolution significantly and at the same time reduce the measuring time cannot be determined for certain at this point.

References for Chap. 5 and 6

Brain

Cerebral metabolic studies in vivo by phosphorous-31 NMR. Prichard JW et al., Proc Natl Acad Sci USA 80: 2748 (1983)
High-resolution proton NMR study of cerebral hypoxia in vivo. Behar KL et al., Proc Natl Acad Sci USA 80: 4945 (1983)
Developmental changes of creatin kinase metabolism in rat brain. Norwood WI et al., Am J Physiol 244: C205 (1983)
In vivo measurement of phosphorous-31 nuclear magnetic resonance spectrum of aging mouse brain. Golczewski JA et al., Physiol Chem Phys 15: 13 (1983)
In vivo flux between phosphocreatine and ATP determined by two-dimensional phosphorous NMR. Balaban RS et al., J Biol Chem 258: 12787 (1983)
Noninvasive observations of fluorinated anesthetics in rabbit brain by fluorine-19 NMR. Wyrwicz AM et al., Science 222: 428 (1983)
^{31}P-NMR studies on the metabolism of high-energy phosphorus compounds in the living rat brain: the effect of halothane anesthesia and hypoxia. Yuasa T et al., Brain Nerve 35: 1089 (1983)

Kidney

Phosphorous-31 nmr studies of energy metabolism in perfused rat kidney. Rhodes RS et al., J Surg Res 35: 373 (1983)
Enhanced recovery of renal ATP with postischemic infusion of ATP-magnesium chloride determined by phosphorous-31 NMR. Siegel NJ et al., Am J Physiol 245: F530 (1983)

Liver

Simultaneous carbon-13 and phosphorous-31 NMR studies of perfused rat liver. Cohen SM, J Biol Chem 258: 14294 (1983)
Application of NMR to the study of liver physiology and disease. Cohen SM, Hepatology (Baltimore) 3: 738 (1983)
Natural abundance carbon-13 nmr spectroscopy of liver and adipose tissue of the living rat. Canioni P, Alger JR, Shulman RG, Biochemistry 22: 4974 (1983)
Structure and metabolism of mammalian liver glycogen monitored by carbon-13 nmr. Sillerud LO, Shulman RG, Biochemistry 22: 1087 (1983)
Phosphorous-31 NMR in the study of liver metabolism in vivo. Quistorff B, Engkagul A, Chance B, Pharmacol Biochem Behav 18: 241 (1983)

Heart

Phosphorous-31 NMR spectroscopy of brain and heart. Ligeti L et al., Adv Exp Med Biol 159: 281 (1983)

Phosphorous-31 NMR studies of enzyme kinetics in perfused hearts from thyroidectomized rats. Seymour AL et al., Biochem Soc Trans 11: 376 (1983)

The temperature-dependence of steady-state creatine kinase fluxes in rat cardiac muscle. Metthews PM et al., Biochem Soc Trans 11: 174 (1983)

The temperature dependence of creatine kinase fluxes in the rat heart. Matthews PM et al., Biochim Biophys Acta 763: 140 (1983)

Nuclear magnetic resonance. Dawson MJ, Card Metab 1983: 309

Phosphorus-31 NMR study of the recovery characteristics of high energy phosphate compounds and intracellular pH after global ischemia in the perfused guinea pig heart. Brooks WM et al., J Mol Cell Cardiol 15: 495 (1983)

In vivo phosphorus-31 NMR studies of myocardial high-energy phosphate metabolism during anoxia and recovery. Neurohr KJ et al., FEBS Lett 159: 207 (1983)

The relationship between global myocardial ischemia, left ventricula function, myocardial redox state, and high energy phosphate profile. A phosphorus-31 NMR study. Whitman G et al., J Surg Res 35: 332 (1983)

Abnormal phosphocreatine metabolism in perfused diabetic hearts. A ^{31}P NMR study. Pieper GM et al., Biochem J 210: 477 (1983)

Observation of a second phosphate pool in the perfused rat heart by phosphorus-31 NMR: is this the mitochondrial phosphate? Garlick PB et al., J Mol Cell Cardiol 15: 855 (1983)

Glycolysis is not activated by pH in the Langendorff heart at pH values above 5.7. Bailey IA et al., Biochem Soc Trans 11: 278 (1983)

Evaluation of high-energy phosphate metabolism during cardioplegic arrest and reperfusion: a phosphorus-31 NMR study. Pernot AC et al., Circulation 67: 1296 (1983)

Nuclear magnetic resonance and positron emission tomography in cerebral vascular disease. Budinger TF, Cerebrovasc Dis 13: 7 (1983)

A protective effect of insulin on reperfusing the ischemic rat heart shown using phosphorus-31 NMR. Seymour AM et al., Biochim Biophys Acta 762: 525 (1983)

Protective effect of nifedipine in myocardial ischemia assesses by phosphorus-31 NMR. Ruigrok TJC et al., Eur Heart J 4C: 109 (1983)

Mathematical analysis of isotope labeling in the citric acid cycle with applications to carbon-13 nmr studies in perfused rat hearts. Chance EM et al., J Biol Chem 258: 13785 (1983)

High-resolution proton NMR studies of perfused rat hearts. Ugurbil K et al., FEBS Lett 167: 73 (1984)

Muscle

A proton-NMR study on lactate and intracellular pH in frog muscle. Seo Y et al., Jpn J Physiol 33: 721 (1983)

Fatigue in retrospect and prospect: phosphorus-31 NMR studies of exercise performance. Chance B et al., Int Ser Sport Sci 13: 859 (1983)

Phosphorus NMR spectroscopy of cat biceps and soleus muscles. Kushmerick MJ et al., Exp Med Biol 159: 303 (1983)

Phosphorus-31 NMR studies of smooth muscle from guinea pig tenia coli. Vogel HJ et al., Biosci Rep 3: 863 (1983)

The effect of exercise on skeletal-muscle intracellular pH in normal and phosphorylase kinase-deficient mice: a phosphorus-31 NMR study. Stevens AN et al., Biochem Soc Trans 11: 92 (1983)

Energetics of smooth muscle tenia cecum of guinea pig: a phosphorus-31 NMR study. Vermue NA, Nicolay K, FEBS Lett 156: 293 (1983)

Bioenergetics of intact human muscle: A phosphorus-31 NMR study. Taylor SJ et al., Mol Biol Med 1: 77 (1983)

Proton NMR of intact muscle at 11 T. Arns C et al., FEBS Lett. 165: 231 (1984)

Tumors

Differences in metabolite levels upon differentiation of intact neuroblastoma and glioma cells observed by proton NMR spectroscopy. Navon G et al., FEBS Lett 162: 320 (1983)

A simple approach for in vivo phosphorus-31 NMR spectral studies of rat tumors. Block RE et al., J Magn Reson 53: 509 (1983)

In vivo phosphorus-31 NMR study of the metabolism of murine mammary 16/C adenocarcinoma and its response to chemotherapy, x-radiation, and hyperthermia. Evanochko WT et al., Proc Natl Acad Sci USA 80: 334 (1983)

Loss of high-energy phosphate following hyperthermia demonstrated by in vivo phosphorus-31 NMR spectroscopy. Lilly MB et al., Cancer Res 44: 633 (1984)

Monitoring response of chemotherapy of intact human tumours by ^{31}P *NMR.* Ross B et al., Lancet 8378: 641 (1984)

Whole-body NMR Spectroscopy

Anatomy and metabolism of the normal human brain studied by magnetic resonance at 1.5 Tesla. Bottomley PA et al., Radiology 150: 441 (1984)

NMR Tomography

Reviews and Books

The Hammersmith clinical experience with NMR. Steiner RE, Clin Radiol 34: 13 (1983)

Clinical prospects of NMR. Worthington BS, Clin Radiol 34: 3 (1983)

NMR imaging: An overview of the physical principles, clinical potential, and interrelationship with radionuclide imaging. Partain CL et al., Nucl Med Ann 1983: 231

Clinical magnetic resonance imaging. Margulis AR, Higgins CB, Kaufman L, Crooks LE (Ed) Radiology Research and Education Foundation, San Francisco 1983

Methodological Progress

Measurement of spin-lattice relaxation times in NMR imaging. Pykett IL et al., Phys Med Biol 28: 723 (1983)

A comparison of the noise characteristics of projection reconstruction and two-dimensional Fourier transformations. Ortendahl DA et al., IEEE Trans Nucl Sci NS 30: 692 (1983)

Real-time NMR clinical imaging in paediatrics. Mansfield P et al., The Lancet 8362: 1281 (1983)

High-Resolution magnetic resonance imaging. Crooks LE et al., Radiology 150: 163 (1984)

Dünnschicht-NMR-Imaging mit einem neuen T_2-gewichteten 3-D-Verfahren. Friedburg H et al., Fortschr Röntgenstr 140: 464 (1984)

Head

Serial NMR imaging in patients with cerebral infarction. Sipponen JT et al., J Comput Ass Tomogr 7: 585 (1983)

NMR imaging of tumors in the posterior fossa. McGinnis BD et al., J Comput Assist Tomogr 7: 575 (1983)

NMR imaging of intracerebral hemorrhage in the acute and resolving phases. Sipponen JT et al., J Comput Assist Tomogr 7: 954 (1983)

NMR evaluation of stroke. Bryan RN et al., Radiology 149: 189 (1983)

Kernspintomogramm eines im CT nur unzureichend dargestellten Germinoms. Stober T, Huber G, Huk W, Fortschr Röntgenstr 139: 648 (1983)

NMR imaging of the posterior fossa: 50 cases. Bydder GM et al., Clin Radiology 34: 173 (1983)

NMR imaging in multiple sclerosis. Lukes SA et al. Ann Neurol 13: 592 (1983)

Anatomy and metabolism of the normal human brain studied by magnetic resonance at 1.5 Tesla. Bottomley PA et al., Radiology 150: 441 (1984)

Cerebral abnormalities: Use of calculated T1 and T2 magnetic resonance images for diagnosis. Mills CM et al., Radiology 150: 87 (1984)

Magnetic resonance imaging of brain tumors: measurement of T1. Tsutomu A et al., Radiology 150: 95 (1984)

Chronic subdural hematoma: demonstration by magnetic resonance. Sipponen JT et al., Radiology 150: 79 (1984)

Head trauma evaluated by magnetic resonance and computed tomography: a comparison. Han JS et al., Radiology 150: 71 (1984)

Magnetic resonance imaging of the orbit: A premilinary experience. Han JS et al., Radiology 150: 755 (1984)

Primary intracranial tumor imaging: A comparison of magnetic resonance and CT. Brant-Zawadski M et al., Radiology 150: 435 (1984)

Ear, Nose, and Throat

Magnetic resonance imaging of the neck: Part I normal anatomy. Stark DD et al., Radiology 150: 447 (1984)

Magnetic resonance imaging of the neck: Part II pathological findings. Stark DD et al., Radiology 150: 455 (1984)

Thorax

NMR of pulmonary arteriovenous fistula: effects of flow. Webb WR et al., J Comput Assist Tomogr 8: 155 (1984)

Cross-sectional imaging (CT, NMR) of branchial cysts: report of three cases. Kreipke DL, Lingemann RE, J Comp Assist Tomogr 8: 114 (1984)

The effect of pulmonary edema on proton NMR relaxation times. Skalina S et al., Invest Radiol 19: 7 (1984)

Multiplane magnetic resonance imaging of the heart and major vessels. Higgins CB et al., AJR 142: 661 (1984)

NMR: Principles of blood flow imaging. Mills CM et al., AJR 142: 165 (1984)

Gated magnetic resonance imaging of congenitial cardiac malformations. Fletcher BD et al., Radiology 150: 137 (1984)

Cardiac imaging using gated magnetic resonance. Lanzer P et al., Radiology 150: 121 (1984)

Volume and planar gated cardiac magnetic resonance imaging: A correlative study of normal anatomy with Thallium-201 SPECT and cadaver sections. Go RT et al., Radiology 150: 129 (1984)

Magnetic resonance imaging of the thorax in childhood. Brasch RC et al., Radiology 150: 463 (1984)

Magnetic resonance imaging of the pericardium: Normal and pathologic findings. Stark DD et al., Radiology 150: 469 (1984)

Magnetic resonance imaging of the breast. El Yousef SJ et al., Radiology 150: 761 (1984)

Multisectional saggital and coronal magnetic resonance imaging of the mediastinum and hila. Webb WR et al., Radiology 150: 475 (1984)

Abdomen and Retroperitoneal Space

NMR imaging of experimentally induced liver disease. Stark DD et al., Radiology 148: 743 (1983)

NMR imaging in the evaluation of the liver: a preliminary experience. Borkowski GP et al., J Comput Assist Tomogr 7: 768 (1983)

Die klinisch-radiologische Bedeutung der verschiedenen Untersuchungsparameter in der NMR-Tomographie des Abdomens. Rupp N et al., Fortschr Röntgenstr 139: 359 (1983)

Coronal NMR imaging of the abdomen at 0.5 Tesla. Kressel HY et al., Radiology J Comput Assist Tomogr 8: 29 (1984)

Magnetic resonance and CT of the normal and diseased pancreas: a comparative study. Stark DD et al., Radiology 150: 153 (1984)
Chronic liver disease: evaluation by magnetic resonance. Stark DD et al., Radiology 150: 149 (1984)
Hepatic tumors: magnetic resonance and CT appearance. Moss AA et al., Radiology 150: 141 (1984)
Magnetic resonance imaging of transfusional hemosiderosis complicating thalassemia major. Brasch RC et al., Radiology 150: 767 (1984)

Others

Fetal imaging by NMR: a study in goats. Foster MA et al., Radiology 149: 193 (1983)
Erfahrungen bei der NMR-Tomographie des Skelettsystems. Reiser M, Rupp N, Stetter E, Fortschr Röntgenstr 139: 365 (1983)

Contrast Media

NMR contrast enhancement study of the gastrointestinal tract of rats and a human volunteer using nontoxic oral iron solutions. Wesbey GE et al., Radiology 149: 175 (1983)
Intravenous chelated Gadolinium as a contrast agent in NMR imaging of cerebral tumours. Bydder GM et al., The Lancet 8375: 485 (1984)
Contrast-enhanced NMR imaging. Brasch RC et al., AJR 142: 625 (1984)
Characteristics of Gadolinium-DTPA complex. Weinmann H et al., AJR 142: 619 (1984)
Intravenous chelated Gadolinium as a contrast agent in NMR imaging of cerebral tumours. Bydder GM et al., Lancet 8375: 485 (1984)

Chemical Shift Imaging

In vivo one-dimensional imaging of phosphorus metabolites by phosphorus-31 NMR. Haselgrove JC et al., Science 220: 1170 (1983)
NMR: In vivo proton chemical shift imaging. Pykett IL, Rosen BR, Radiology 149: 197 (1983)
Spatial mapping of the chemical shift in NMR. Mansfield P, J Phys D 16: L235 (1983)
Three-dimensional Fourier transform NMR imaging. High resolution chemical-shift-resolved planar imaging. Hall LD et al., J Magn Reson 56: 314 (1984)
NMR imaging/spectroscopy system to study both anatomy and metabolism. Bottomley PA et al., Lancet 2 (8344): 273 (1983)
Spatially resolved high resolution spectroscopy by „fourdimensional" NMR. Maudsley AA et al., J Magn Reson 51: 147 (1983)

7 Subject Index

abdomen, NMR tomogram 92–97, 99
absorbed energy 14
acetyl-CoA 34
adenosinediphosphate (ADP) 35, 36, 43
adenosinetriphosphate (ATP) 2, 3, 10, 11, 29, 30, 34–46, 88
ADP (s. adenosinediphophate) advantage, NMR tomography 58, 105
alanine 31–34
–, transamination 32, 33
alignments, antiparallel 14, 61, 110
–, parallel 14, 61, 110
alloys, nonferromagnetic 101
α-glucose 31
alternating field, B_1 107
amalgams 101
amylose 21
analog to digital converter 27, 28
arginine, brain tissue 29, 31
arrythmias 99
astrocytoma, grade II 81
atomic nuclei, magnetic properties 4, 62
atoms, hydrogen 9, 51–53, 67
ATP (s. adenosinetriphosphate)

back projection 57f.
– – reconstruction, filtered 58
background noise 13
bar magnet 4
barium 84
–, contrast medium 58
β-glucose 31
bibliography, Chap. 1–3 47–50
–, (references) Chap. 4 102–104
biomolocules 85
biopsy material 46f.
blood flowing 82, 99
– plasma, fluorinated 88
– pressure 99
Bohr magneton 85
Boltzmann's constant 110
–, distribution 110, 112
brain studies 44, 45
breast carcinoma 72, 76
B_1-alternating field 107
B_1-field 61, 62
B_1-partial wave 107

carbon 6, 13, 88
–, atom 30–33
–, dioxide 34, 35
carcinoma diagnosis 76
cardiac pacemaker 27, 100, 101, 105
cardioplegic agent 40f.
Carr-Purcell experiment 71, 110
– –, (multiple spin-echo) technique 86, 87
– –, sequence 71
– –, spin-echo, multiple 86, 87
carrageenan 86
cerebral abscess 90, 91
– infarction 91, 92
chemical binding 110
– shift 9–12, 15, 36–38, 42, 110, 111
cholangiocarcinoma 96
choline, brain tissue 29, 31
CH_3OH (methanol) 10
circumferentially polarized partial waves 107
cirrhosis 94
–, alcohol-induced 94
–, ascites 94
citric acid cycle 33–35
^{13}C-labeled glucose 88
clips, hemostatic 101
–, surgical 100, 101
^{13}C-NMR signal 12
– spectroscopy 29
– spectrum 21
coils
–, cryomagnetic 111
–, field profiling 43
–, geometry 43
–, radiofrequency 18, 27, 31, 43, 54, 110
–, receiver 18, 63–65, 79, 108
–, shim 27, 112
–, surface 43–47
–, transmitter 61, 107
computer tomographic scan 90–92
computed tomography 51, 54
concentration changes 58
contrast material 86
– medium, barium 58
– –, concentration 86
– –, selective 86
coordinate system, inverse transformation 108
– –, rotating 61, 107–109

– transformation 108
costs 47, 88, 105
creatine 42, 43
–, phospho- (PCr) 2, 3, 29, 30, 34–46, 88
– phosphokinase 35–38
cross-sectional image 53, 54, 57
– –, water-filled capillary tubes 53, 54
– – plot 60
cryomagnet 110, 112
CT/NMR comparison 90
current density 100
cutaway drawing, superconducting magnet 23
cyst, proencephalic 46

damage, irreversible 39
–, ischaemic tissue 47
data-acquisition process 82
decarboxylation 34
decay curve 19, 27, 110
– –, summation 21
delay time 70
– – T_1 65
– – T_E 73, 78
– – T_R 73
δ-scale 10
demyelinating process 92
dental appliances, steel 101
detection, phase-sensitive 108
deuterium 88
diagnostic NMR system 25
diphosphoglycerate 2, 3, 11
disadvantage, NMR technique 58, 105
D_2O (heavy water) 88
drosophilia melanogaster 99
double resonance 112

ECG 93, 97, 99
echodensity 45
electric currents, body 100
– moment 111
electromagnet 21–23
electromagnetic spectrum 8, 9
electronic integration 10
electrons 112
–, shell 9
–, surrounding nucleus 110
–, unpaired 84
energy absorbtion 18
– difference 6, 7
– levels 4–6, 14, 15, 18, 62, 110, 111, 113
enzyme activities 36, 99
– deficiency diseases 3, 34, 105
– – states 47
– deficit 39
equilibrium constant 42
– value 64
evaporation 110

excitation pulse 58, 59, 62, 63, 68, 70, 71, 82, 86, 110, 111
– technique, multiplanar 72
exponential function 18

$FADH_2$ molecules 35
Fast-Fourier transformation 110
fibrillation, ventricular 100
field, B_1 61, 62
– gradient 53, 57, 101
– profiling 43
– – coils 43
– strength 7, 16, 22, 24, 47, 51
– –, nominal 21
FID (s. free induction decay) fluid, fast-moving 82, 83
–, slow-moving 82, 83
–, stationary 82
fluorine 88
flow 85
– velocities 82, 83
– vessels 51
flowing blood 82
– fluid, time diagram 83
Fourier, Jean Baptiste Joseph 20
– transformation 19, 21, 27, 28, 56, 63, 110
– –, one-dimensional 59, 60
– –, two-dimensional 58–61
free induction decay (FID) 19, 58–60, 63, 65, 67–71, 82, 110
– – – –, summation 67, 69
frequency axis 53

gadolinum DTPA 84
– –, dimeglumin 85, 86
γ-phosphorus atom 42
gauss 110
genetic injury 99
– studies 99
gluconeogenesis 32
glucose 32, 33
– ^{13}C-labeled 88
– 1-phosphate 35
– signals 34
–, synthesized, liver 32
glycolysis 36–39
glycerol-3-phosphorycholine 45
– 3-phosphorylethanolamine 45
glycogen phosphorylase 34, 38
gold 101
gradients 58f., 100
–, direction 57
–, fields 53, 57, 101
–, G_x- 52, 59–61
–, G_y- 52
–, G_z- 52, 59–61
–, magnetic field- 52, 53, 55, 57–59, 88, 99

gradients, x- 56
-, z- 56
gravitational field 106
GTP (guanosine triphosphate) 35

harm, irreversible 100
head examinations 8
health effects 9
- hazard 100
- risks 99, 100
heart rate 99, 101
heavy atoms (barium/iodine) 84
- water (D_2O) 88
helium 22–24, 111
- evaporation 23
hemostatic clips 101
hepatitis chronic 95
hepatoma 94
^{1}H-NMR spectrum 10, 29
homogeneous magnetic field 51, 52, 56, 65, 68, 99, 111
Hounsfield scale 73
hydrogen 87, 88
- atoms 9, 51–53, 67
- tissue water 51
- ion concentration 37
- NMR signal 13
- nuclei 4, 6, 10, 12, 15, 16, 53, 54, 57
- - concentration 54
hypothetical spectrum 9
hypothyroid patient 41

image affect, tissue properties 73
- contrast 66, 71, 72, 80, 105
-, cross sectional 53, 54, 57
-, gray-scale 58
- intensity 91
- plane 54, 56, 57, 73, 82, 87
- -, multiple sections 73
- -, sagittal 55
- -, selecting 56
-, proton-density 71, 90
- quality 70, 86, 87
- reconstruction 54f.
- -, two-dimensional 59
- spin-echo 72
imaging methode 90
- parameter 17, 51, 78, 80
- -, image contrast 80
- -, pertinent 77
- -, tissue properties 80
- - T_E 78, 79
- - T_R 78, 79
-, soft-tissue 105
- technique 51
- -, multiplanar, time diagram 73
- time 86, 87, 105

- whole-body 88
implants, cardiac pacemaker 105
-, metallic 105
-, risks 100
-, surgical 101
individual magnetization 61, 62
- projections 57
- scans 13
- spectra 13, 60
injurious effects 100
integration, electronic 10
interval T_R 15
intervertebral discs 92
intrinsic motion 89
- rotation, nuclei 111
inverse transformation 108
inversion recovery (IR) experiment 65f.
- - image 65–67, 78, 91
- - measurement 65
- - NMR tomogram 92
- - sequence 65, 78, 80, 91
- - signal 65
- - technique 63, 66, 91
- - -, time diagram 79
inverting pulse 58, 63, 65–69, 71, 79, 110, 112
investigations, biological 29
-, medical 29
iodine 84
- content, thyroid 88
ionizing radiation 3, 51
- X-ray CT 100
ions, hydrogen, concentration 37
-, metal 85
-, paramagnetic 84, 85
IR (s. inversion recovery) irradiation, radio-frequency 100
-, short-wave 100
ischemia 38, 39
isolated nucleus 9
isotope abundance 7
-, NMR-active 6, 87
ISU-unit 111, 112

jaundice, obstructive 96, 97

kilogauss (kG) 110

lactic acid 35, 38–40
Larmor frequency 107, 111
- relation 7, 9, 52, 107
lattice 14, 15, 112
light-sensitive photoreceptors, retina 100
line width 13
liver glycogen 34
longitudinal magnetization 112
- relaxation time 63
- - - T_1 63
Lorentzian line 111

macromolecules 74
macroscopic scale 61
magnet, individual 61
–, superconducting 21–25, 105, 112
magnetic field 5, 24, 43, 44, 113
– –, auxiliary 55, 57
– – axis 106, 111
– – break down 24
– – changing 99, 100
– – collapse 24
– – direction 106, 107
– –, external 14, 106, 111
– – gradient 52, 53, 55, 57–59, 88, 99
– – –, linear 56
– – –, time varying 101
– –, homogeneous 51, 52, 56, 65, 68, 69, 99, 111
– –, –, static 111
– – intensity 24
– – lines 26
– –, main homogeneous 52
– –, nonhomogeneous 52, 58, 67–69, 112
– –, nonuniform 67
– –, profiling 45
– –, static 99–101
– –, – varying 100
– – strength 6–9, 16, 24, 52, 53, 56, 77, 78, 101, 105, 110–112
– – –, nonhomogeneous 112
– – –, T 112
– –, time-varying 100
– – uniformity 21, 22, 24
– –, whole-body system 24
– induction 7
– moment 84, 85
– phosphenes 100
– properties 4, 6, 7, 61, 62, 111, 112
– –, atomic nuclei 62
– radiofrequency field, coupling 113
– screen 9
magnetization, individual 62
–, net- 61–63
–, – M_o 61, 62
magnetogyric ratio 7, 52, 111
magnitude, field strength variations 17
materials, ferromagnetic 101
–, nonferromagnetic 101
Mc Ardle syndrome 38, 39
metabolic phophorus compounds 35
metabolites 89
metal ions 85
metallic implants 105
– –, risks 100
metals, ferromagnetic 101
methanol (CH_3OH) 10
methylene, hydrogen atoms 29
–, tissue fat 30

modulation 55, 57
molecular symmetry 34
multiplanar technique 72, 86
multiple sclerosis 92
– spin-echo technique 71, 86, 110
muscle samples 36

NAD^+ 34, 35, 38, 39
NADH 34, 38
–, molecules 35
^{23}Na-NMR tomogram 88
neoplasms, diagnostic 105
net magnetization 61–64, 107, 108
– –, M_o 61, 62
neutralizing coils (shim coils) 27, 112
neutrons 6
newborn infants 3
nickel chloride, paramagnetic 84
nitrogen 23
– (^{14}N) 88
NMR-active atom 9
– – isotope 6, 87
– nucleus 10, 87, 112
–, advantages 105
– contrast media 84f.
– /CT comparison 90
–, disadvantages 105
– energy level 111
– experiment 4, 7, 14, 108, 111
– –, pulsed 63
– –, rotating coordinate system 106
– frequency shifts 10
– imaging 51, 54, 82, 90, 101
– – intensity 73
– – system 89
– – techniques 52
– – time efficiency 70
– investigations 89
– phenomenon 51
– plane selection 54
– properties, tissue 80
– pulse experiment 19
– relevant tissue properties 51
– –, (flow velocity) 51
– – –, (relaxation times) 51
– – – –, (water content) 51
– resonance frequencies 8
– sensitivity 51, 87, 88
– signal 5, 6, 10, 11, 13, 28, 43, 83, 89, 111, 112
– – acquisition 18
– –, ^{13}C- 12
– –, ^{31}P- 12
– spectra, conventional 88
– –, high-resolution 9
– –, individual 88
– –, summation 21
– spectrometer 112

– –, conventional 43
– –, topical 43
– spectrometry, whole-body 47
– spectroscopic examinations, organs 43
– spectroscopy 6, 10, 29, 89, 99, 105
– –, brain 105
– –, ^{13}C 29
– –, ^1H 29
– –, phosphorus 2
– –, ^{31}P 34, 39, 41, 46, 47
– spectrum 6, 10, 15, 113
– –, ^1H 10
– , one-dimensional 59, 60
– –, ^{31}P 10, 11
– –, signal intensity 54
– –, two-dimensional 59, 60
– system, block diagram 27
– –, resolution 10
– –, typical layout 25
– –, whole-body 105
– technique 89
– –, advantages 58
– –, disadvantages 58
– –, medical applications 1
– –, two-dimensional 58
– tomogram 73 f., 90–92
– –, abdomen 82
– –, coronary 17
– –, dissecting aortic aneurysm 98
– –, head 65–67, 89, 90
– –, –, axial 1, 89
– –, –, mediosagittal 2
– –, –, sagittal 89
– –, heart 97
– –, ^{23}NA 88
– –, ^{31}P 88, 89
– –, retroperitoneum 17
– –, spin-echo 71
– –, thorax 54
– –, T_1 65, 66
– tomographic experiment, time diagram 70
– – system 51
– tomography 47, 51, 89, 100, 105
– –, abdominal organs 92–99
– –, advantage 58
– –, aneurysms 93
– –, biological effects 100
– –, data strategies 70
– –, elements, other than hydrogen 87 f.
– –, first applications 89
– –, head 72
– –, health hazard 100
– –, – risks 99, 100
– –, heart 93
– –, physical basis 52
– –, principle 53
– –, spine 92

– –, thoracic 92, 93
– –, – cavity 98
– zeugmatography 51
nominal field strength 21
nuclear population, thermal equilibrium 111
– quadropolar moment 111
– resonance 7
– spin 106, 111
– –, procession 106, 107
nuclei, NMR-active 87, 112
– orientation, spatial 106
nucleus, hydrogen 4, 6
–, NMR-active 10
– specific constant 9
null plane 55, 57

object plane 56
opposing magnetic field 24
oxaloacetic acid 32–34
oxidative phospharylation 35
oxygen, molecular 86

pancreatic carcinoma 97
paramagnetic compounds 84, 86
– ions 84, 85
– nickel chloride 84
parameter T_E, T_R and T_I 80
partial wave 107, 108
–, B_1-field 107, 108
P_b-relaxation time 74
PCR (s. phosphocreatine)
P_f-relaxation-time 74
ph, intracellular 2, 36–39
–, intramuscular 39
phase 59
– sensitive detection 108
phosphate, adenosindi- (ADP) 35, 36, 43
–, adenosintri- (ATP) 2, 3, 29, 30, 34–46, 88
–, guanosintri- (GTP) 35
–, inorganic (Pi) 2, 3, 29, 30, 34, 36, 37, 40, 41, 43–46, 88
–, ribose-5- 45
–, sugar- (SP) 36, 37
phosphocreatine (PCr) 2, 3, 12, 29, 30, 34–46, 88
–, kinase (pK) 36, 38
–, resonance 36
phosphodiester 41
phosphoenolpyruvic acid 32
phosphorus atoms 3, 10, 11, 30, 42
– compounds 34
– –, metabolic 35
– concentration 88
– metabolites 2, 3, 34, 36, 41, 46, 88, 105
– NMR spectroscopy 2
– nucleus-31 34
– (^{31}P) 88

phosphorylation, oxidative 38
phosphorylcholine, glycerin-3- 45
phosphoryletanolamine, glycerol-3- 45
physical principles 51
physiologic effects 99
– –, parameters 100
P_i (s. phophate inorganic)
  piperdine-N-oxyl 84
pK (s. creatine phosphokinase)
plane, image 54, 56, 57, 73, 82, 87
–, null- 55, 57
–, objet- 56
–, selection- 54, 82
–, x- 57, 60, 61
–, y-z- 61
–, z- 57, 60, 61
–, z-y- 108
^{31}P-NMR signal 12
– spectroscopy 34, 37, 39, 41, 46, 47
– spectrum 10, 11, 36–39
– tomogram 88, 89
– –, signals 88
polyglucose 34
population difference 5, 6, 14, 61, 70, 71,
  86
precautions 25, 28
precession 61, 62, 106, 107, 108, 111
precessional angle 61
– frequency 111
– –, nuclei 107
pregnant women 3
profiled field 45, 47
projection, back- 57 f.
–, individual 57
–, single 57
–, x- 57
–, z- 57
proton density 69, 70
– – image 71, 90
protons 4, 6, 12, 112
–, water 30
pulse 18, 61, 108, 110
–, excitation- 58, 59, 62, 63, 68, 70, 71, 82, 86,
  110, 111
–, inverting- 58, 63, 65–69, 71, 110, 112
–, radiofrequency- 19, 27, 55, 63, 107
–, read- 63, 64, 65, 67, 79, 111
–, selective- 56, 73
– sequence 112
– –, saturation recovery 63
–, M- 111
–, M/2 111
pyrrolidine-N-oxyl 84
pyruvic acid 32–35, 38, 39

quadropolar nuclei 111
quantum-mechanical standpoint 106

quantum theory 62
quenching 111

radiation 100
radicals, stable free 84, 85
radiofrequency coil 18, 27, 31, 43, 54, 110
– damping material 28
– energy 18, 19, 42, 52, 61, 74, 100
– field 100, 112
– interference 28
–, irradiating 99
–, magnetic field coupling 113
– pulse 19, 27, 55, 63, 107
– radiation 7, 99, 101
–, physiological effect 99
– reception 18
– transmission 18
radiography, contrast media 84
read pulse 63–65, 67, 79, 111
receiver coil 18, 63–65, 79, 108
relaxation 13, 14, 63
– mechanisms 108
– –, tissue 74
– process 14, 64, 79, 111
– –, T_1 15, 65
– time 16, 17, 67, 83, 84, 90, 91
– –, P_b 74
– –, P_f 74
– –, longitudinal 63
– –, –, T_1 63
– – measurement 60
– –, sensitive 75
– –, spin-lattice 63
– –, –, T_1 63
– –, spin-spin 68, 81
– –, tissue water 84
– –, transverse 63
– –, –, T_2 63
– –, T_1 15–18, 60, 61, 63, 64, 66, 74–78, 82, 84,
  85, 91, 111, 112
– –, T_1-spin-lattice 73, 112
– –, T_2 18, 60, 61, 63, 64, 68, 69, 72, 74, 76–78,
  88, 111, 112
– –, T_2-spin-spin 65, 73, 112
– –, T_2 67
– – in vessels 51
repetition time 17, 86
– –, T_R 14, 63, 78, 81
respiratory chain 34, 35, 38
resolution 105
–, local 87
resonance frequency 6–10, 17, 19, 39, 41, 42,
  51–53, 56, 58, 59, 67, 68, 111
– –, energy 14
– –, nucleus 110
retina, light-sensitive photoreceptors 100
ribose-5-phosphate 45

rotating coordinate system 106
– direction 107
rotational frequency 107

safety coils 24
– devices 24
– recommentations 101
saturation 6, 14, 15, 18, 111
– recovery experiment 15
– – pulse sequence 63
– – technique 111
– transfer 41, 42
scintigraphy 105
sectional image 54
selected plane 54, 82
selective excidation 55, 56
– pulse 56, 73
sequence, inversion recovery (iR) 65
–, pulse 112
–,–, saturation recovery 63
shell electron 9
shift, chemical 9–12, 15, 36–38, 42, 110, 111
shim coils 27, 112
shielding 110, 111
short-wave irradiation 100
signal-acquisition time 70
– (brightness) difference, tissues 80
– decay 68
– fat 31
– height 112
– intensity 13–16, 19, 53, 57, 70, 78
– spectrum 54
– splitting 112
– to-noise ratio 12, 15, 21, 87, 112
–, water 31
single projection 57
skin effect 7, 88, 112
S/N ratio 112
sodium (^{23}Na) 88
– succinate 13
soft-tissue imaging 105
solids 17
sonographic demonstration 46
SP (s. sugar phosphates)
spectral quality 13
spectroscopy, TMR (topical magnetic resonance) 113
spectrum, electromagnetic 9
–, hypothetical 9
–, individual 60
–, total 56
–, two-dimensional 59, 60
spin 4, 112
– decoupling 112
– density 112
– echo 58, 68, 112
– – experiment 68, 69

– – image 72
– – intensity 68, 69
– –, multiple 71
– –, –, (Carr-Purcell) 86, 87
– – NMR tomogram, axial 71
– – technique, multiple 71
– – tomogram, axial 70
– imaging 51
– mapping 51
– quantum number 112
– lattice relaxation time T_1 14–16, 63, 70, 73, 112
– spin coupling 112
– – relaxation 16, 19, 68, 81
– – – time T_2 16, 18, 65, 73, 112
stainless steel 100
standard imaging paramters 81
– signal 10
stray field 26
succinic qcid 34
superconducting magnet 21–25, 112
surface coils 43–47, 105
surgical clips 100, 101
– implants 101
sugar phosphates (SP) 36, 37
syndrome McArdle 38, 39

T_E-delay time 73, 78
T_E-imaging parameter 78, 79
T_E-time 59, 60, 61
tesla (T) 22, 110, 112
tetramethylsilane (TMS) 12
thalassemia major 11
thermal equilibrium 82, 112
thoracic cavity 98
time constant 111, 112
– diagram and image intensity 79
tissue damage, ischemic 3
– discrimination 17, 65, 71, 77
–, fast-growing 76
– fat 29, 31
–, malignant 76
– heating 99, 100
– parameters 78
– properties 73, 79, 80, 105
– –, image affect 73
– –, NMR imaging 78
– water 29, 31, 51
– – content 73–77
– – distribution 1
– – relaxation time 84
titanium 100
titration curve 38
T-magnetic field strength 112
TMR (s. topical magnetic resonance)
TMS (s. tetramethylsilane)
tomogram, water content 81

tomographic experiment 67
– technique, development 51
topical megnetic resonance (TMR) 2, 43,
    113
toatal sensitivity 87
– spectrum 56
tracer 88
transamination 33
transition metal ions 84
– – –, biologic properties 85
– – –, magnetic properties 85
transmitter coil 61, 107
transverse magnetization 112
– relaxation time 63
– – –, T_2 63
T_R-delay time 73
T_R-imaging parameter 78, 79
T_R-interval 15
T_R-repetition time 78
T_R-value 15
t-time 59, 60, 61
T_1-NMR tomogram 65, 66
T_1-relaxation process 15, 65
T_1-relaxation time 15–18, 63, 64, 66, 74–78, 82,
    84, 85, 91, 111, 112
T_1-spin-lattice relaxation time 16, 73, 112
T_1-value 15
T_2-relaxation time 18, 60, 61, 63, 64, 68, 69, 72,
    74, 76–78, 88, 111
T_2-spin-spin relaxation time 16, 18, 73,
    112
T_2-relaxation time 67

ultrasound examination 45

ventilation 86
visual disturbances 100
voltage 99

water, bound 74, 75
–,–, relaxation 75
– content 79, 80, 83, 84, 90
– –, edema 91
– –, tissue 73–76
– – tomogram 81
– – (vessels) 51
– distribution 60
– filled capillary tubes, cross-sectional image
    53, 54
–, free 74, 75
–, relaxation 75
– signal 52
waiting time 70
whole-body examinations 8
– – imaging 88
– – investigation 21
– – magnet system 24, 25, 47
– – spectroscopy, measurements human
    patients 29
– – NMR system 21, 25, 47, 105

x-gradient 56
x-plane 57, 60, 61
x-projection 57
X-ray 51, 54
– –, absorbtion coefficient 73
– – CT 105
– – –, ionizing radiation 100

y-z-plane 61
Zeeman effect 113
zeugmatography 113
z-gradient 56
z-plane 57, 60, 61
z-projection 57
z-y-plane 108

G. Gademann

NMR-Tomography of the Normal Brain

1984. 68 figures. Approx. 115 pages
ISBN 3-540-13233-3

Contents: Physical and Technical Fundamentals. – Contrast Behaviour. – Normal Anatomy of the Head As Seen in the NMR Image. – Sagittal Sections, Frontal Sections. – Horizontal Sections. – References. – Subject Index.

Here at last is the guide to NMR visualization that busy clinicians and researchers have been waiting for. It includes both a brief introduction to the physicial and technical aspects of NMR as well as an atlas of NMR scans using the spin-echo and inversion-recovery techniques, which demonstrate normal head anatomy in three different planes.

The author shows that NMR offers several distinct advantages over CT scans: It opens up new visualization possibilities beyond conventional horizontal sections, such as sagittal and frontal sections parallel to the plane of the face, giving unfamiliar but particularly clear images of the head and brain. NMR does not require radioactive contrast media, thus avoiding injurious effects on biological systems. And, since bone generates a very low NMR signal, the NMR tomogram produces an astonishingly clear image of the posterior cranial fossa without the bony artefacts that plague the CT scan.

The excellence and astounding variety of images that NMR allows in the head, brain, face and neck will make this book of interest not only to neurologists and neurosurgeons, but also ENT specialists and oral surgeons, for whom NMR imaging is destined to become an important addition to their diagnostic possibilities.

Springer-Verlag
Berlin
Heidelberg
New York
Tokyo

Frontiers in European Radiology

Editors-in-Chief:
A. L. Baert, E. Boijsen,
W. A. Fuchs,
F. H. W. Heuck

Frontiers in European Radiology aims
- to provide a platform for European radiological research
- to strengthen the position of radiology in clinical medicine
  by integrating the various approaches to clinical and
  experimental radiology

Frontiers in European Radiology (FER) addresses radiologists all over the world with the goal of improving the international exchange of information on all aspects of radiological research. This exchange has unfortunately been limited in the past, especially by the language barriers involved. As a result, Europe's contribution to scientific progress in this interdisciplinary specialty has influenced only regional developments. A first step toward rectifying the situation was taken in Hamburg in September 1979, where the formation of the association of European University Radiologists was discussed and decided upon.
FER is the logical continuation of that initiative; it provides a forum for scientists in European clinical and experimental radiology where important reports on progress in the field can be presented in a depth not possible in a journal. It is a concise source of detailed information for those wanting to keep abreast of the scientific progress in this field.
FER includes both short, important communications and detailed treatments relating to the results of research. All contributions appear in English. The contributions are of high quality and concerned solely with progress in the field of radiology. Unsolicited papers are considered for publication.
Detailed information for authors is available upon request from the editors and Springer-Verlag.

Volume 4

1984. 82 figures in 144 separate illustrations.
Approx. 150 pages. ISBN 3-540-13410-7

Volume 3

1984. 80 figures in 143 separate illustrations.
III, 136 pages. ISBN 3-540-11446-7

Volume 2

1982. 70 figures in 84 separate illustrations.
V, 103 pages. ISBN 3-540-11349-5

Volume 1

1982. 113 figures in 187 separate illustrations.
V, 170 pages. ISBN 3-540-10753-3

Springer-Verlag
Berlin
Heidelberg
New York
Tokyo